ACCIDENTAL GENIUS

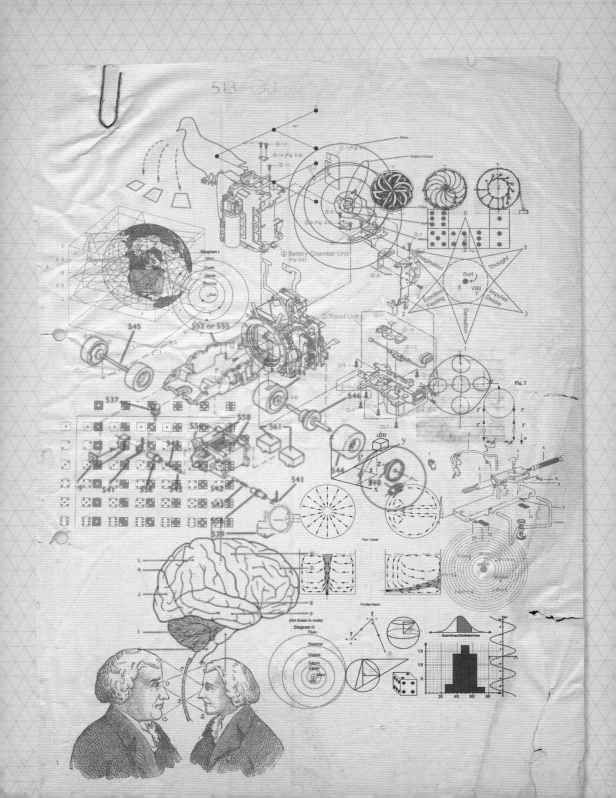

ACCIDENTAL GENIUS

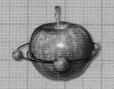

Richard Gaughan

METRO BOOKS

NEW YORK

Project Editor: Asha Savjani
Editorial Assistant: Holly Willsher
Designer: Jason Anscomb
Art Director: Michael Charles
Managing Editor: Donna Gregory
Publisher: James Tavendale

Metro Books
122 Fifth Avenue
New York, NY 10011

ISBN: 978-1-4351-2557-5

Printed and bound in China

1 3 5 7 9 10 8 6 4 2

CONTENTS

THE EUREKA MOMENT

★

Once upon a time . . .

INTRODUCTION

A king wants to make an offering to the gods; so he hires a smith to fashion a golden crown. But rumors reach the king: the goldsmith has been known to mix less expensive silver with gold. The king is enraged, but doesn't know how to test the allegations. Luckily, he knows of a brilliant young man whose insights have already astounded the community. The king asks the young man for help. Unfortunately, the young man can think of no way to determine if the offering has been adulterated; he goes to the baths to ease his frustrations. He lowers himself into a full tub. He sits for a moment, enjoying the warmth of the waters, but his relaxation is disturbed by the sound of the water dripping over the side of the tub.

Suddenly he leaps out of the tub and runs to the king (not bothering to dry off—in fact, not bothering to dress himself). "I've found it! I've found it!" he shouts as he cavorts through the streets. The king was named Hiero, the city was Syracuse on the island of Sicily, and the young man, of course, was Archimedes. Being Greek, Archimedes was not shouting the English words "I've found it," but the Greek word "Eureka!" He realized that by testing how much water the "gold" offering displaced he could tell if silver had been intermingled.

This story of Archimedes has survived more than two thousand years, even though it probably isn't true. (It was first written almost two hundred years after it would have happened, and Archimedes, who wrote extensively on the properties of buoyancy and density, never mentioned the incident.) The story survives because we can imagine it to be true. We have experienced the excitement of discovery, and we know it can transport us into such a state of elation that we forget such things as hunger or exhaustion (although, thankfully, we don't usually forget our clothes).

French engraving of Archimedes (c.287–212 BCE), Greek mathematician and inventor. Andre Thevet, Les Vrais Portraits Et Vies Des Hommes Illustres, *1584.*

Even though it's not true, Archimedes's "Eureka Moment" remains the model for the flash of inspiration behind scientific discovery and technological innovation.

Scientific progress is achieved through a combination of slow, steady progress and sudden, rapid breakthroughs. Experimental scientists plan their tests, carry them out, and interpret their results meticulously. Theoretical scientists rigorously develop their proofs and derive their mathematical models. These are the skills necessary for slow and steady progress, responsible for most of our understanding of our world. These same scientists also have flashes of insight, some moment where an unusual measurement or an equation of an unexpected form suddenly just makes sense. That flash of insight then leads to another round of painstakingly careful work and the cycle repeats. Sometimes the flashes of insight rise directly out of the fertile ground prepared by

Thomas Edison (1847–1931), US inventor. Photographed working in his chemical laboratory in West Orange, New Jersey, c.1905.

repetitive routine, but sometimes there is another spark, a chance occurrence that triggers an extraordinary moment of piercing clarity. When the moment is recognized and acted upon, the world can change in virtually an instant. But if the moment is not recognized, the potential dies.

What makes the difference? What determines the fate of an accidental observation or chance circumstance? What factors determine which yield fruit and which are ignored?

First, there's preparation. Imagine this: a group of researchers are busy mixing a batch of chemicals to produce a gas. They know exactly what is supposed to happen. A certain ratio of ingredients at a certain temperature for a certain time will give them exactly what they need. They also know the properties of a wide range of materials and how they can be used. They are prepared to notice if something doesn't turn out the way it should.

The second ingredient, the obvious in a chance discovery, is the chance occurrence itself: opportunity. For our chemists, something goes wrong and the mixture doesn't turn out to be a gas, instead they end up with a fine white powder, something like shredded coconut.

Our team of chemists can now do one of two things: they can toss their mistake into the trash or they can observe that the product that was accidentally created has unique and interesting properties. The majority of people, including the majority of scientists, will be so focussed on what they were trying to achieve that they won't see what they weren't looking for. But every once in a while, preparation and opportunity come together and the accident is recognized as something unique.

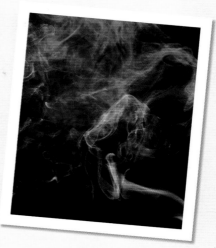

Experiment is an essential part of the scientific process; this is when many accidental discoveries have been made.

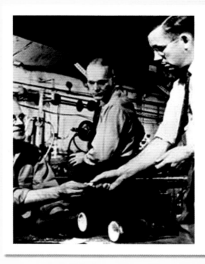

Dr. Roy Plunkett (right) reenacts discovery of DuPont™ Teflon® PTFE in his laboratory.

But the story is not complete. There is a third critical element: desire.

Let's go back to our chemists. Instead of a container of gas, they have a fine white powder. Their job was to create a gas. They failed, and they recognized that in their failure they had created something that hadn't been seen before. Now what? They could just say, "that's funny," and go back to work. They could put their new chemical in a jar and resolve to come back to it later. Or they could try to figure out the properties of the new substance they've made and learn how it had come together.

Again, scientists and engineers are like everyone else. They have a job to do and they know that distractions can keep them from being successful. So there's a temptation to stay on track, to stay focussed on the task at hand, to resist the urge to follow curiosity. It's only when the compulsion to discover is stronger than the need to show results that the accidental discovery comes to fruition. Our chemists, for example, took their work completely off track when they found their unexpected white powder. They investigated the properties of their new compound and found a material that was extremely corrosion resistant, electrically insulating, and very slippery...a perfect coating for surfaces subject to friction and corrosion: polytetrafluoroethylene (PTFE). You might know it better by its trade name, DuPont™ Teflon®.

PTFE was discovered only because Roy Plunkett, a scientist with DuPont, had the preparation, opportunity, and desire to see the discovery for what it was. All the stories of chance discoveries that follow contain those three elements: preparation, opportunity, and desire.

Sometimes scientists are reluctant to acknowledge the role of luck in their discoveries. They quite rightly argue that without education and training, without careful planning and organization, without detailed observation and incisive interpretation, no discovery would ever be made.

All of that is true. But scientific discovery is like so many other elements of our own lives. For things to work out just right many different factors come into play. Some of these factors are under our control, and some are not.

An analogy can be found in sporting events. How many times have two closely matched teams met in a championship game? Each team has opportunities to take the lead, but the match is tied and goes to extra time. The errant bounce of a ball, a misstep on a slick surface, a burst of crowd noise that hampers communication; and one team is crowned champion. No one would deny that luck can be a factor. On the other hand, if the team hadn't trained, didn't have the skill, hadn't developed a game plan then they would have had no chance at all. And, of course, if fatigue or doubt outweigh desire, then the effort fails. Preparation, opportunity, and desire all must be present.

Maybe one reason scientists tend to downplay the role of chance is that it tends to diminish the importance of their own insight, their intelligence, their creativity—in short, their preparation. Often, later in their careers when their reputations are secure and their genius acknowledged, they find it easier to acknowledge the role of random

Clinton Joseph Davisson, US physicist, 1881–1958.

The celebrated Davisson-Germer experiment confirmed the de Broglie hypothesis that particles of matter have a wave-like nature, which is a central tenet of quantum mechanics.

chance. For example, in 1926 Clinton Davisson discovered (or, perhaps more accurately, verified) that electrons—tiny solid particles—could act as if they were quite wispy waves, like a beam of light. When he first reported his experiments in 1927 he started his paper with the phrase, "In a series of experiments now in progress, we are directing a narrow beam of electrons against a target...," with not a word about why he and Lester Germer were choosing to make such an experiment. Ten years later, when the validity of his striking results was accepted and he was accepting his Nobel prize in Physics, he told a different story. "The true story contains less of perspicacity and more of chance. The work actually began in 1919 with [an] accidental discovery...And then chance again intervened; it was discovered, purely by accident, that [electrons bounced like waves]."

On the other hand, you get situations like that of J. Robin Warren, recipient of the 2005 Nobel Prize in Physiology or Medicine. He was acknowledged for his work uncovering the role of the bacterium *Helicobacter pylori* in causing duodenal ulcers. At several stages in his work he got "lucky." That is, circumstances came together to help lead him to his discovery. One piece of the puzzle only came to light when a bacterial sample was left to incubate over the five-day-long Easter Holiday, instead of only the usual

48 hour incubation. Without that step, which showed the presence of the relatively slow-growing Heliobactor pylori, his discovery might have been delayed for who knows how long. But when he accepted his award he said,

> Obviously, as with any new discovery, there is an element of luck, but I think my main luck was in finding something so important. I think the best term is serendipity; I was in the right place at the right time and I had the right interests and skills to do more than just pass it by.

J. Robin Warren received the 2005 Nobel Prize in Physiology or Medicine.

Warren was right to be a little sensitive about claims that he was just "lucky." His work is a perfect example of the lone scientist who develops a theory that flies in the face of accepted fact. If it were not for his perseverance, his painstaking and rigorous search for evidence, the logical edifice he built from mountains of collected data, it is likely the scientific world would still not accept the role of bacteria in gastric disease.

Warren uses the word "serendipity" to describe the role of chance in his discovery. It's worth taking a look at that term.

In 1754 Horace Walpole, an English man of letters, found it necessary to come up with a word to describe a new type of discovery. While doing something completely unrelated, he ran across some information that was exactly what he needed for some research he was doing. This seemed very similar to what he had read in a children's story about princes from the Isle of Serendip, who "by accident and sagacity" made discoveries they were not searching for. He called this kind of discovery "serendipity." Since then the word has seen increasing usage as people acknowledge the role of "accident" in scientific discovery.

There are other ways in which chance reckons into scientific discovery. To truly be serendipity, the discovery needs to be unsought. A researcher

FACING PAGE:
*Archimedes (257–212
BCE), mathematician,
geometrician, physicist,
and technician of
antiquity is shown here
caught in the act of
accidental discovery.*

can be looking for something fruitlessly when chance intervenes to provide the key to revealing the very answer that was being sought. And sometimes chance shows itself by changing researchers' circumstances, setting them on the path that seemingly leads inevitably to the discovery history remembers them for.

This book presents stories of moments when fate stepped in to stimulate a moment of sudden insight that changed human understanding. These moments could happen to any of us—Archimedes was not the first man to spill the bathwater—but it takes more than just random chance to create a Eureka Moment. Archimedes saw the water spill, and he was prepared to see the implications because he knew that a volume of gold was heavier than an identical volume of silver. And, whether it sprang from the need to help his friend or to build his reputation, Archimedes had the desire to determine if his flash of insight would prove true.

The anecdotes here are separated in chronological order, divided roughly into scientific eras—times within which the nature of scientific investigation had a unique character. The distinctions are a bit arbitrary, but still useful. We start in a time before science was a well-defined process, in the years prior to 1800. The next set of stories comes from the years between 1800 and 1900, when scientific investigation was still largely driven by individual endeavor and supplied with individual resources. From 1900 to 1940 science was becoming more internationalized, with frequent scientific conferences helping to stimulate discovery. From 1940 to 1980 research continued its trend toward specialization, driving scientists into smaller niches, but worldwide

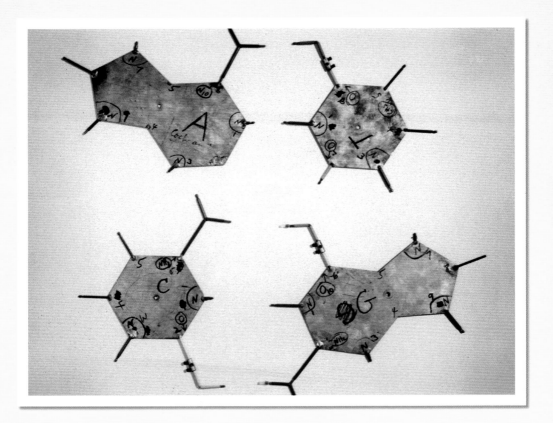

Aluminum templates as part of Watson and Crick's model representing the structure of DNA.

FACING PAGE: *Isaac Newton's discovery of gravity is one of many serendipitous moments in science.*

communication helped broaden perspectives as well. And the years from 1980 to the present have seen science grow as a "corporate" activity, both within research and development laboratories associated with companies, and with large review boards determining which projects should be publicly funded.

Whatever the nature of the era, scientific progress has pushed forward with slow, churning activity, then leaped ahead with sudden insights. In these stories chance, accident, fate, luck—whatever you choose to call it—has stepped in to create opportunities to advance human knowledge. In each of these eras there have been limitations, barriers to inquiry and discovery. Although the job of "doing science" has changed, human nature has not. When chance presents itself, human curiosity must triumph over constraints for the promise of discovery to be realized.

Preparation, opportunity, and desire met to create the initial Eureka Moment, and those three elements have been present at every serendipitous discovery since then. This collection of stories will take us on a tour of human discovery, an expedition through a set of unique moments of concentrated experience and chance occurrences that have built upon each other to create our understanding of the world.

MYTHIC MOMENTS: APOCRYPHAL APPLES

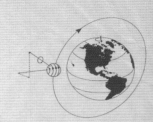

Scientific discovery did not sit around and wait for "science." Far back in human pre-history, science and technology still advanced—although we might not recognize them by those names. And wherever science and technology advanced, it is very likely chance played a role. It's hard to imagine that chance did not play a role in the development of the first wheel or the first intentionally started fire, but those stories are so far buried in the past that any discussion of what might have happened would be pure speculation. The stories in this section are not quite that speculative.

But they're close.

They're apocryphal. The dictionary definition says that means they're of dubious authorship and questionable authenticity. Doesn't mean they aren't true, just that it would be difficult to prove their truth.

Some accidental discoveries are just barely within the range of recorded history; one of these stories comes from that range. But all the stories come from a time before there were any "scientists." Science in this era was just another philosophical discipline, and the people who tried to broaden and deepen our understanding of the world in which we live were called "natural philosophers." They were no less careful with their work, but they were working without the structures that "scientists" of later eras would take for granted. In fact, their work would help define the process of science. As in every other time of history, accidents of fate played a role in their discoveries.

BAKING SODA INTO GLASS
An unnamed Phoenician, *c.*3000 BCE

UNLOCKING THE SECRETS OF LIGHT
Abu Ali al-Hasan Ibn Al-Haitham reveals secrets of optics, *c.*1011

IT'S ALL DOWNHILL FROM HERE
Isaac Newton discovers Universal Gravitation, 1666

A SPARK OF INSPIRATION
Pieter van Musschenbroek and Ewald Jürgen von Kleist discover the Capacitor, 1745–46

A JUMP SPARKS A NEW UNDERSTANDING
Luigi Galvani discovers "Animal Electricity," 1781

AN UNNAMED PHOENICIAN

*c.*3000 BCE

BAKING SODA INTO GLASS

★

IN THE MEDITERRANEAN *world prior to the rise of the Greek city-states and well before the centuries of Roman rule, the Phoenicians sailed the waters of the sea. Sailing in those days was one of the riskiest endeavors: lose sight of land and there was no instrument for navigating other than intuition. So sailors would hug the shore, putting in at night to recover from the rigors of a trying day.*

THE ORIGIN OF GLASS

On one particular night, a group put in on a sandy beach and built a fire ring with blocks of minerals they were carrying. They were amazed to see the sand and minerals melting into a red-hot glowing stream of liquid. The next morning they saw a hard, transparent solid where the previous night's liquid had flowed. For the first time, people had created glass.

Thousands of years later Pliny the Elder, in his treatise *The Natural History*, told the story of the origin of glass. Interestingly enough, he tells the tale in the section called "The Natural History of Stones":

The story is, that a ship, laden with niter, being moored upon this spot, the merchants, while preparing their repast upon the sea-shore, finding no stones at hand for supporting their cauldrons, employed for the purpose some lumps of niter which they had taken from the vessel. Upon its being subjected to the action of the fire, in combination with the sand of the sea-shore, they beheld transparent streams flowing forth of a liquid hitherto unknown: this, it is said, was the origin of glass.

The color of clarity

When we think of "glass" we think of the clear drinking vessels or the windowpanes we look through, but scientists define glass as a material that behaves as a solid but has a structure that makes it look more like a really thick liquid. Although common glasses are primarily made from silica sand mixed with soda and lime, other materials can make glass. The properties of the glass of course depend upon the materials that go into making it. The color, for example, changes when small amounts of additional chemicals are added to the mix.

The Romans spread glass technology and raw materials through their empire. When the empire collapsed, glassmaking also declined. A thousand years later the technology was revived, but raw materials were difficult to come by. In seventeenth-century England, for example, potash refined from wood ash replaced mineral soda. Glassmaker George Ravenscroft added some extra lead oxide and created a shiny strong glass that came to be known as lead crystal. Ravenscroft didn't record the role of chance in his discovery.

FACING PAGE LEFT: *Pliny the Elder (Gaius Plinius Secundus) was a Roman scientist and scholar.*

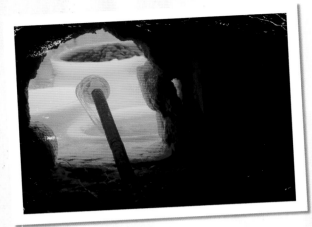

Glass furnaces are natural gas or fuel oil-fired, and operate at temperatures up to 2867°F (1575°C).

A HAPPY ACCIDENT

Most glass today is made primarily of silicon dioxide, also called silica usually found in nature, in the form of sand. Glass in the ancient world was also made mostly from sand, but silicon dioxide melts at a very high temperature, around 3092°F (1700°C). That temperature is much hotter than fires in the ancient world could reach. But when soda—sodium carbonate, called "niter" by Pliny—is added to silica, it reduces the melting point to about 2552°F (1400°C). That's a temperature that could be reached in ancient times. According to Pliny's account, happy accident led to the discovery of glass manufacturing.

Even if the story isn't strictly true (and 5000 years later it's impossible to tell), it's highly likely that accident played a big part in the discovery of glass. Perhaps the process was discovered during the manufacture of faience, an earthenware fabrication technique that ends up with a final product that is part clay and part a glassy glaze. If the proportions are a bit off and the temperature is a little hotter, a translucent glass is the result. It's easy to imagine a simple mistake: a little bit of extra soda falls into a crucible, the fire is fanned a bit too briskly, and Bob's your uncle: glass!

It's difficult to imagine circumstances surrounding the discovery of glass where chance didn't play a part. Certainly someone would have run across a fulgurite, a glass nodule created when lightning hits a sandy beach. But it would have been impossible to duplicate the amount of heat delivered by the lightning; so any formal developmental program would

When lightning strikes sandy soil, the air and moisture present in soil are rapidly heated, and the resultant explosion-like expansion forms a fulgurite.

MORT DE PLINE L'ANCIEN.

have had to wait for the happy coincidence of a soda deposit and sand being present during a raging fire. But even though luck was a critical component of this discovery, it took far more than simple chance to turn it into a method for manufacturing glass.

DELIBERATE ATTEMPTS

What Pliny didn't discuss was the frenzy of activity that must have followed such a chance event. If the effect had been dismissed as a mere curiosity, there wouldn't have been deliberate attempts to recreate and refine the glass manufacturing process. And without refinement, glass never would have been more than an idle curiosity, because although silica melts at a lower temperature when mixed with sodium carbonate, the resulting glass gradually dissolves when it gets wet. So someone needed to discover that mixing lime (calcium carbonate) with the silica and soda makes the glass durable!

So, whether it was the unnamed Phoenician of Pliny's story or an Egyptian craftworker, somebody observed the formation of glass and had the desire to develop the process. A chance observation triggered a development with clear advantages to human history.

In his eagerness to observe the eruption of Vesuvius, Pliny the Elder approaches dangerously near and is killed by falling rocks, a martyr to science.

Workers round the furnace in Mr. Apsley Pellatt's glass house in Holland Street, Blackfriars, London, 1842.

ABU ALI AL-HASAN IBN AL-HAITHAM

REVEALS SECRETS OF OPTICS

c.1011

UNLOCKING THE SECRETS OF LIGHT

ABU ALI AL-HASAN IBN AL-HAITHAM

(or Alhacen, in Latinized form) was born in Basra, Persia in 965 CE. He was groomed for a lucrative career following in his father's footsteps as an upper-level civil servant. Instead, in the isolation of his own house, he experimented with the nature of light and vision, advancing science and——more important——establishing an experimental process for evaluating ideas about the nature of reality. Yet he never would have been in a position to seed the growth of science if fate had not pushed him into an unlucky circumstance: he was imprisoned for ten years.

UNLUCKY CIRCUMSTANCE

Alhacen started his career as a civil servant under the Abbasid dynasty. But he found the work boring. Also, because the political hierarchy was also a religious hierarchy, he was expected to spend much of his time in religious study, but he soon realized that none of the competing religious interpretations could lay claim to "truth." So he moved himself into another field: he studied natural philosophy and mathematics. He applied his abilities to practical problems as well as esoteric questions, and he rapidly gained acclaim as a respected man of learning. So it was that when

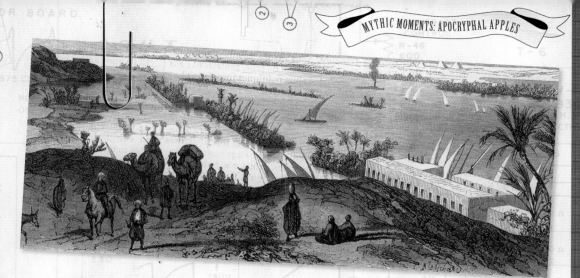

The River Nile breaks its banks.

he read reports of the devastation caused when the Nile River flooded each year, and he saw the maps that had been drawn, he suggested that it might be possible to dam the Nile River. Egypt had just recently been conquered by the Fatimid Caliphate, political and religious rivals of the Abassid Dynasty that ruled Persia. The Caliph al-Hakim bi-Amr Allah, who ruled in the new city of Cairo, heard of Alhacen's claim and hired him to do the job. When Alhacen got to the site—the right spot, incidentally, where the Aswan dam was built one thousand years later—he found that the impression he'd received from the maps and reports was in error. With the technology available to him, there was no way the river could be dammed successfully.

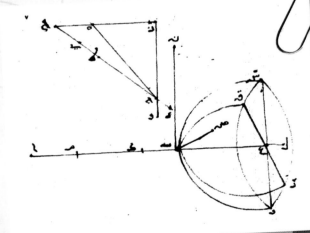

If another caliph had ruled in Cairo, that might not have been a problem. But al-Hakim was notorious for not taking bad news well. Random executions were not unknown under his rule, and men had been killed for disappointing al-Hakim in less significant ways than failing to dam the Nile River. Alhacen took steps to preserve himself. By the time he returned from the upper Nile back to Cairo, he was acting like a madman. Because of this madness, Alhacen did not have the intention

Diagram from manuscript on optics by Abu Ali al-Hasan Ibn al-Haitham, using mathematical methods, 965–1039 AD.

to cause trouble; so under Islamic Law al-Hakim could not have him punished. But he could keep him from causing more trouble by imprisoning Alhacen in his own house. Enforced solitude pushed Alhacen to apply his curiosity and incisive mind to problems he could address in a limited space, with restricted resources.

He turned his thoughts to the problem of vision. The ancients had wondered how it was possible that images of the world around us make their way into our minds. Some thought objects in the external world sent little copies through space and into our eyes. Some thought our eyes sent rays out to the objects around us. They thought and thought and thought. Perhaps Alhacen remembered that just "thinking" about damming the Nile made him think it could be done, until physical observation made him realize it was impossible. Perhaps he just looked at the house that had become his prison and wanted to make it as interesting as possible. But whatever the source, it seems to be clear that Alhacen was the first to say that observation and experiment were more important than just philosophical musings. That is, if philosophical musings clash with experimental results, experimental results must prevail. With that as his guiding principle, Alhacen set to learning about the nature of light.

He blocked light from entering a room in his house, except through one tiny opening. He set up five lanterns outside the room and observed five separate lantern images inside. He blocked the direct line of one lantern to the "pinhole" in the wall, and the image of that one lantern disappeared. He set up strings and tubes to show that light traveled in straight lines. He put those experiments together with some more everyday observations: First, when the eyes are opened, both nearby objects and distant stars are perceived simultaneously. Second, staring at bright objects hurts the eyes and leaves afterimages. He realized his darkened room was a model for the eye. Light comes into our eyes through a small opening, entering in straight lines. Light either originates with an object (the sun, a candle) or it bounces off an object (trees, books), but it all obeys the same rules when it travels to our eyes.

THE GROWTH OF SCIENCE

Alhacen went on to study the bending of light by curved surfaces and by the atmosphere, the mathematical nature of reflection, and the optics and psychology underlying optical illusions. These studies jump-started the growth of science on many fronts, a fortunate circumstance started through an accident of fate that Alhacen could only have regarded as unfortunate.

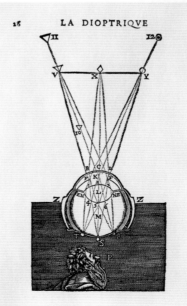

LA DIOPTRIQVE

Line engraving showing "Optics" from the nineteenth century.

Spreading the news

Communication in the world at the end of the first millennium was more or less at a walking pace. Still, there was trade across political, religious, geographical, language, and cultural boundaries. Foodstuffs, fabrics, and finery may have been the foundation of trade, but there was also an exchange of ideas. As the centuries passed, Europe's thirst for ideas brought the works of Alhacen into an arena ripe for a spark.

Alhacen's Kitab Al-Manazir, or Book of Optics, provided that spark. Alhacen's experimental model of inquiry needed only slight refinement to become the scientific method that we know of today as the most efficient way to learn about the world in which we live. Specifics of his optics also seeded the investigations of other luminaries of scientific history such as Roger Bacon and Johannes Kepler. If only all forced confinement led to such discoveries!

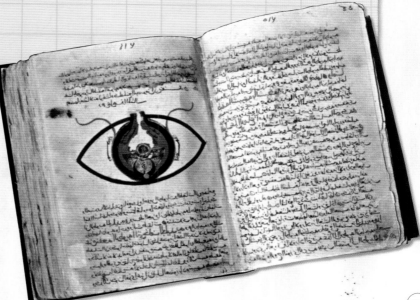

Two pages from thirteenth-century Arabic manuscript of Hunayn's Ten Treatises on the Eye.

ISAAC NEWTON

DISCOVERS UNIVERSAL GRAVITATION

1666

IT'S ALL DOWNHILL FROM HERE

23-YEAR-OLD *Isaac Newton rested underneath an apple tree at his family's Woolsthorpe Manor home. An apple fell upon his head. After he said "ouch" he wondered why it was that the apple fell straight down from the branch on a line toward the center of the Earth. He looked up at the quarter-moon in the sky and suddenly he realized: whatever was responsible for pulling the apple to the ground was the same thing that was responsible for making the moon travel around the Earth. The Law of Universal Gravitation was discovered.*

OR SO THE STORY GOES

Two things we know. First, Newton himself said he mused on the nature of gravitational attraction while watching apples fall from the trees in his gardens. Second, he would not have been home

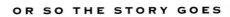

at Woolsthorpe if not for the plague sweeping through Europe. And if he hadn't been musing about apples in his gardens, who knows where his thoughts might have wandered?

Newton graduated from Cambridge University in 1665 and came home for a brief rest, but before he could return to his studies at Cambridge the University was closed because of the plague. As Newton himself noted, "in those days I was in my prime of age for invention, and minded mathematics and philosophy more than at any time since." But rather than being immersed in the lively academic world of Cambridge, he was stuck in a quiet backwater village with no intellectual companionship except for the sun and the apples that fell in the yard. For most of two years the plague forced him to turn his curiosity on sunlight and apples. His observations led him to change the nature of human understanding of our world.

Newton had significant insight into so many different fields it would be difficult to even list them here, but the one most closely associated with accident is his discovery of the law of gravity. Before Newton, people thought of the heavens and Earth as different realms, operating under different rules. But Newton, sitting at his window, saw the apples fall. No matter how high the tree is, he noticed, the apple still falls to the ground.

Woolsthorpe Manor in Lincolnshire, the birthplace of Isaac Newton (1642–1727).

Newton uses a prism to decompose light.

If a particular apple tree is twice as tall, four times as tall, eight times as tall, we still expect the apple to fall straight to the ground. No matter how tall the tree is, we would be surprised if the apple didn't fall straight down. Even if an apple tree stretched all the way to the moon, we'd still expect the apple to fall straight down...so doesn't it make sense that whatever force is pulling the apple down is also pulling on the moon?

THE MATH

People already knew how far away the moon was, and how fast orbited around the Earth; and they also knew the moon was staying up there. That is, the moon is as far away from the Earth at the end of an hour as it is at the beginning. Newton put those things together to figure out how much the Earth's surface dropped away from the moon in that hour, which must be the same as how far the moon "falls" toward the Earth in that hour. The answer showed the moon was pulled by one-thirty-six-hundredth ($1/3600$) of the force that pulled an apple to the ground. But wait! If an apple on a really tall tree and the moon are both being pulled by the same thing, shouldn't they feel the same amount of force?

Newton went back to the math. He knew that the distance from the center of the Earth to the center of the moon was 60 times larger than the distance from the center of the Earth to the center of the apple. If it was truly the same force, as he believed it to be, then it must vary inversely with the square of the distance between them. That is, objects twice as far from each other will be subject to a force one fourth as large.

Issac Newton's cradle is a device that demonstrates conservation of momentum and energy.

He also thought of it another way

Imagine a cannon fired straight out from the top of a mountain. The cannonball would start falling as soon as it was fired, but it would also be shooting away from the mountain; so it would land some distance away. If the powder in the cannon were doubled, the cannonball trajectory would be flatter and the cannonball would travel further away. At some point there would be enough force behind the cannonball so that the landing point could not be determined without considering the curvature of the Earth. The cannonball falls about 16 ft (5 m) in the first second. The Earth's surface curves down and drops away about 16 ft (5 m) every 8 km (5 m); so if the cannonball is shot fast enough to travel 8 km (5 m) in one second, it is as high above the Earth's surface after one second as it is at the start. Since (assuming it is above the air where there's no friction) it is still traveling as fast the next second and the second after that, it never comes down: the cannonball is orbiting the Earth—just like the moon!

Measuring and transporting cannon balls in the eighteenth century.

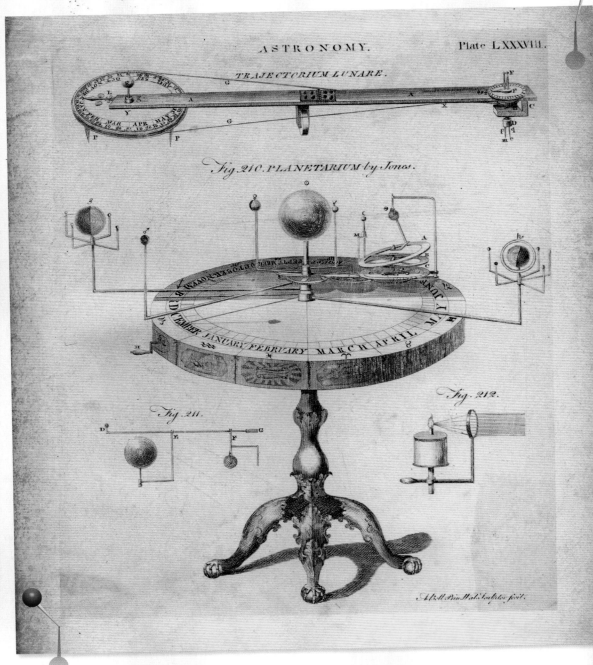

Astronomical objects, namely a trajectorium lunare and an orrery or planetarium by William Jones, 1797.

THE PRINCIPIA

This is the Law of Universal Gravitation, that the force of gravitational attraction is the same in the heavens as it is on Earth. By showing this, Newton demonstrated that the universe fits into a framework, and if we can learn how it works here on Earth, we can learn how it works everywhere. By mathematically connecting the motion of a falling fruit to the motion of the whirling planets he philosophically connected heaven and Earth. In 1687 he finally published his book describing these insights, *Philosophiae naturalis principia mathematica*, or the *Principia* for short, and the world was never the same.

If chance hadn't forced him to his family orchard, he might have turned his talents to other problems, leaving the world waiting for the understanding that physical rules apply all throughout the universe.

IRASCIBILITY AS A DRIVING FORCE

Certainly chance played a role in Newton's major discoveries. The plague was sweeping through England, disrupting any kind of activity that required people to assemble. Commerce, entertainment, diplomacy, and education all suffered as quarantine and isolation were used to slow the spread of the disease that killed millions of people in outbreaks over hundreds of years. Without the plague, Newton wouldn't have been forced into isolation with his thoughts and his orchard. But would that have kept the discoveries from being made?

NEWTON VS HOOKE

There is one element of Newton's character that makes that an open question: Newton was a cantankerous fellow.

Natural philosophers of the time were desperately trying to explain the world around themselves, including the motion of the planets. Newton was not the only one. When Robert Hooke, one of England's leading scientists, claimed to have discovered a law governing gravitation, Newton said, in effect, "that's no explanation; I've come up with better ideas years ago looking at my orchard." When Hooke said, "prove it," Newton couldn't resist the urge to shove it in his face. So it was that he further refined and formalized his insights into the *Principia* more than 20 years later. So perhaps Newton's combative nature was as much responsible as his genius, along with the chance that forced him into two years of independent contemplation and experiment.

ABOVE LEFT: *Robert Hooke, English scientist. Author of* Micrographia *(1665), in which he published results of his microscopic investigations.*
ABOVE RIGHT: *Isaac Newton, engraving from 1856.*

PIETER VAN MUSSCHENBROEK
AND EWALD JÜRGEN VON
KLEIST

DISCOVER THE CAPACITOR

(1745—46)

*Pieter van Musschenbroek,
Dutch mathematician and
physicist, discovers the principle
of the Leyden jar (known also as
Kleistian jar), 1746.*

A SPARK OF INSPIRATION

IN THE EIGHTEENTH CENTURY, *shocking people
was the fashion. Not the kind of shock delivered when a teenager comes
home with their hair cut into an arm-length orange mohawk, but a physical
jolt of electricity. Demonstrations of static electricity were seen in the
homes of the elite. One of the favorite demonstrations was to suspend a boy
from the ceiling by silken threads, charge him up with static electricity,
then have him touch a person standing on the floor, sending a spark from
finger to cheek. (Perhaps that's why teenagers today get their revenge by
surprising mom with an orange mohawk.) Investigators thought there was
an "electrical fluid" that pervaded everything. Attempting to capture this
fluid, two natural philosophers independently developed a device which we
would now call a capacitor.*

Fig. 44.—Wimshurst's Machine.

Playing with fire—electrical fire. This electrostatic generator was made by James Wimshurst (1832–1903). One of the parlor tricks that "electricians" would use to entertain at parties involved isolating a young woman from electrical connection with the earth and connecting her to a "friction machine." She would collect static electricity, which would be discharged, sparking through the fascinated men who would take the challenge.

THE ELECTRIC KISS

Static electricity had been known of since the time of the ancient Greeks, but it wasn't until the middle of the seventeenth century that machines were invented to generate static electricity. Natural philosophers would work these machines (which one way or another rubbed two materials against each other—such as wool and glass) and touch one hand to either side. The strength of the electricity was measured by how strongly the experimenter was shocked! And, once the spark was drawn, it was gone. How could they study something that wouldn't stick around?

Ewald Jürgen von Kleist, an amateur natural philosopher living in the Prussian province of Pomerania, was experimenting with electricity. He was probably trying to "catch" electrical fluid. He connected a static electricity machine to a nail that was immersed in water, the water being held in a small jar. When he held the jar with one hand, then touched the nail with his other, he received a shock which stunned his arms and shoulders. He thought he was catching electrical fluid, but the only reason the water held charge is because his hand was there and he was connected to the Earth. When the static electricity machine sent charge to the nail, it completed a circuit through von Kleist's hand to the Earth, leaving the jar holding electrical charge.

BURGUN

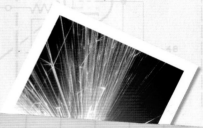

Pieter van Musschenbroek, Dutch mathematician and physicist, performs electrical experiments at Leiden, where he was a professor.

A random spark of inspiration?

Von Kleist experimented a bit, and found that he could hold the jar, connected to a static electricity machine, but then disconnect the machine, walk to another room holding his electrified jar, and the jar would stay strongly electrified. He reported this to leading scientists of the day, but, either because his description was incomplete and hard to follow or because he had no scientific credentials, his results were ignored. Luckily for the scientific world, a few hundred miles away someone else was making the same chance discovery.

Pieter van Musschenbroek, Dutch mathematician and physicist, 1692–1761.

After his experiments, van Musschenbroek reported to the scientific community:

"I would like to tell you about a new but terrible experiment, which I advise you never to try yourself, nor would I, who experienced it and survived by the grace of God, do it again for all the kingdom of France.

This warning did exactly what similar warnings do for those teenagers mentioned at the start of this story: soon people everywhere were doing such experiments."

Pieter van Musschenbroek was a Professor in Leyden in the Netherlands. His friend Andreas Cunaeus, a lawyer, spent time in Musschenbroek's well-equipped laboratory. Musschenbroek was trying to get an electrical spark from water. He was an experienced electrician, as such investigators were called at that time, so when he hooked a static electricity machine to a wire in a glass jar full of water on his bench he made sure the jar was electrically insulated. Cunaeus, an amateur, was not aware of the importance of electrical insulation; so he simply picked up the jar as it was charging. Then he later touched the wire leading into the jar and received a jolt of electricity.

THE LEYDEN JAR

The apparatus—a jar with a metallic wire, nail, or chain inserted into a liquid inside—became known, after Musschenbroek's town, as a Leyden Jar. It has the property of storing electrical energy, of making it available for study, and it was the first step toward developing our modern understanding of electricity. It was thought of as a jar in which to capture electrical fluid, but later experiments found that the water, the nail, even the jar itself, were not necessary to store electrical energy. The electrical energy is not contained in the jar, but it is held between a conductor on the inside and a conductor on the outside. Without the grounded human hands to create the conductor on the outside, the electrical circuit could never have been complete. So two different researchers made identical mistakes. They thought they could catch electricity in a jar and they—by chance and by accident—held the jar in their hands.

The Leyden Jar is now known as the charge-holding component called a capacitor, used in just about every electronic device ever produced.

The Leyden Jar, a late eighteenth century device for storing static electricity, consists of two conductors separated by an insulator.

Crashing ahead

At the time of the invention of the Leyden Jar little was known of the nature of electricity. We now know that electrical current is carried by tiny particles that we call electrons and that electrical current operates through the same physical principles, no matter the source. But one of the controversies of the mid-eighteenth century concerned the difference between the "natural" electricity of lightning and the "artificial" electricity from static electricity-generating friction machines.

Benjamin Franklin was trying to answer that question. He and his son flew a kite in a thunderstorm, capturing charge in a Leyden Jar. Experiments with this charged Leyden Jar showed that the captured "natural" electricity behaved the same as captured "artificial" electricity. When Franklin attempted to get his results published by the Royal Society in England, he was ignored. Franklin then described an experiment for a "sentry box" in which a person could hold on to one end of an iron rod during a thunderstorm. Some say this was Franklin's joke——daring someone to try to challenge the power of electricity, and the power of Franklin's ideas. That idea, which may have been conceived facetiously, led to the development of the lightning rod.

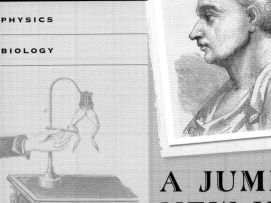

Fig. 403.

A JUMP SPARKS A NEW UNDERSTANDING

★

ABOVE, BELOW, AND RIGHT:
Galvani's frog experiment, 1780, observing contractions in a frog's legs when brought into contact with an electrical machine.

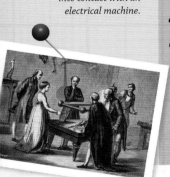

ANIMAL ELECTRICITY *In the middle of the eighteenth century, researchers were asking big questions like "What is the nature of electricity?" and "How can living things move?" In the city of Bologna a professor of medicine embarked on a series of experiments designed to answer both questions at the same time. Although he meticulously planned all of his experiments, chance still played a big role.*

THE INITIATOR

Luigi Galvani was wondering what triggered a leg muscle to contract. And where did all the energy in that leg kick come from? What if, he thought, the mysterious force within living bodies was electricity? So he collected machines that generated and stored static electricity, then he connected them to the nerves of severed frog legs. When the electrical

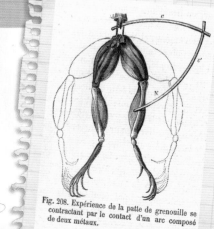

Fig. 208. Expérience de la patte de grenouille se contractant par le contact d'un arc composé de deux métaux.

machines sent a spark through the frog legs, the legs gave a kick! He had discovered that nerves carried electricity to make muscles work. But there was still a big question about the electricity that made muscles move: did it come from the air, like lightning, or did it come from inside the body itself? Then chance stepped in.

A random spark of inspiration?

Galvani wrote his famous paper on animal electricity more than ten years after he began his experiments. Like most scientific papers, he organized his report in a way that would make it easy to understand, not necessarily the way it really happened. He wrote that an assistant just happened to be creating a spark while another was touching the frog's nerve with a scalpel. But his notebooks of the time tell a different story.

He was in the midst of a series of experiments that led him to believe there was an "electric fluid" pervading the air, accumulating in the nerves. While his wife (who often assisted in his experiments) or another assistant elicited a spark from the machine, he touched the nerves and observed the legs jump. That is, at the time, the experiment appeared to be part of a plan, but years later he wrote that it was chance.

The general tendency in scientific reporting today is to downplay the role of chance, and highlight the planning and insight necessary for a successful scientific discovery. Galvani's report seems to go against this trend. The only thing certain is that, as in so many other discoveries, both perception and providence played a part.

ABOVE: *Luigi Galvani's production of an electric current between two metals and a frog's legs. Engraving from Luigi Galvani's De Viribus Electricitatis, 1791.*

When he was writing about it ten years afterward, Galvani explained that:

> *I placed the frog on the same table where was an electrical machine, but the animal was completely separated from it and at a considerable distance from the machine conductor. When one of my assistants, by chance, lightly applied the point of a scalpel to the inner crural nerves of the frog, suddenly all the muscles of the limbs were seen to contract that they appeared to have fallen into violent tonic convulsions.*

CHANCE OBSERVATION

The machine generated a spark at one end of a table, and the frog legs jumped at the other end of the table—without being connected to the machine at all! A chance observation gave Galvani the opportunity, then his desire motivated him to start a whole new series of experiments. At the end the conclusion was clear. The energy to move, Galvani thought, had to come from within the frog legs themselves, energy in the form of "animal electricity." The conclusion was clear, but it was wrong.

In fact, Galvani had made probably the first observation of radio transmission. The machine generating a spark sent out a pulse of radio energy, and the frog's leg and the attached wires were an antenna. Galvani had the opportunity and the desire to understand his observations, but not the preparation. He wasn't to blame: nobody was yet prepared. Radio waves were flying through space, but the groundwork for understanding that idea was 100 years away.

GALVANI VS VOLTA

But Galvani's conclusions sparked a controversy. Alessandro Volta, a physicist working in Pavia, disagreed with Galvani. Volta thought that the source of the energy came from outside the frog's legs. Volta's experiments led him to develop the "pile," the first chemical battery. The battery brought electricity under the control of researchers, and started a whole new era of scientific investigation. Volta was proved right, and Galvani was wrong. Electricity was generated by different metals brought into contact with each other, and not from within living creatures.

Because Galvani had been proved "wrong," many of his conclusions were discarded. It took about 100 years for his experiments to be reconsidered. It turns out Galvani was partially right. Nerves and muscles do work with electrical signals.

His chance observation led to the harnessing of electricity and the understanding that living bodies are biophysical machines. Galvani initiated two whole new fields of study, all from a single spark.

Alessandro Volta (1745–1827) Italian scientist, notable for his invention of the voltaic pile—the first electric battery.

Galvani discovers the existence of animal electricity in an experiment on a recently killed frog, from Le Journal de la Jeunesse, *(Paris, 1880).*

THE AGE OF REASON: THE EMERGENCE OF MYSTERY

★

Between 1800 and 1900 the nature of scientific investigation was changing. Many "basic" discoveries had been made; so someone who wanted to push the frontiers of knowledge needed to become at least somewhat specialized. In the late 1700s, Isaac Newton made his greatest strides in what we would call physics, but he invested equal energy in mathematics, chemistry, alchemy, theology, and even biblical studies. Although he spread his efforts over so many fields, he made great leaps of knowledge that lay the foundation for bringing the physical world into the realm of human understanding. That changed in the nineteenth century.

The "natural philosopher" was disappearing, to be replaced by the "scientist." Structures for mentoring and educating scientists began to be introduced. Active researchers took on the responsibility for guiding "apprentices"; graduate students formally studied under experienced scientists—even the word "scientist" first appeared in this century. Scientific disciplines became separated: physics, chemistry, astronomy, and other specialisms all required increasing focus.

This century also saw increasing awareness that science was important for human progress. The industrial revolution was bringing new goods to a wider range of people than ever before in human history. Scientific research that used to be regarded as an esoteric, impractical endeavor was now being seen as the crucible of innovation. In addition to the importance placed on science in academia, this century also saw the emergence of the corporate research and development laboratory. For example, at his facilities in New Jersey, Thomas Edison assembled teams of engineers, technicians, machinists, and scientists all dedicated to turning technology into marketable products.

The new emphasis placed on science and the new structures established to streamline scientific education didn't change the basic model for scientific investigation. Reduced to its basics, the process begins with an idea about how the world works, then continues with an experiment to test if the idea is correct. There is plenty of room for surprises, and the surprises came. Still, preparation, opportunity, and desire remained the keys to success. But the preparation became more specialized. It's easy to understand why. If, for example, you pour two liquids into a flask, heat them up, and get a black sludge, you need to have enough background in chemistry to know if the result is normal or unexpected. So it was that the accidental discoveries made in a specific discipline in this era tended to be made by specialists in that specific discipline; that is, chemical discoveries were made by chemists.
In this environment where there was increasing belief that the world was understood, every surprising result served as a reminder that increasing knowledge leads to greater mystery. Here are stories of mysteries solved.

THE FIRST "MISTER CLEAN" IS NOT SO WELCOME
Ignaz Semmelweis reduces the incidence of childbed fever, 1847

A NEW WORLD STARTS WITH A DYE
William Henry Perkin invents the first artificial dye, 1856

SIDESTEPPING FATE
Alfred Nobel invents dynamite, 1867

CHICKEN SOUP GONE BAD—GOOD NEWS FOR HEALTH
Louis Pasteur develops chicken cholera immunization, 1879

JELLING THE FUTURE OF BIOLOGICAL RESEARCH
Angelina Hesse develops a bacterial growth medium, 1881

SEEING SKELETONS
Wilhelm Conrad Roentgen discovers X-rays, 1895

ETIENNE-LOUIS MALUS

DISCOVERS LIGHT POLARIZATION
BY REFLECTION

1808

SEEING FROM A DIFFERENT ANGLE

IN THE BEGINNING YEARS *of the nineteenth century, no one understood the nature of light. It was different from anything else that had been studied. Light carried energy and information from one place to another. But what was it? Was it a miniature stream of particles, something like a chain of one rock being tossed after another? Or was it a "vibration," like a wave in water, carried along in something we can't see? A young French engineer, Etienne-Louis Malus, was searching for an explanation for the strange behavior of some crystals. These crystals, called birefringent crystals, create two images. Looking through them, one can see two of everything. But Malus, idly looking through one of the crystals, noticed that the reflection of sunlight in the windows of a neighboring palace only showed one image. He had discovered a new property of light that was to provide a key piece of the puzzle.*

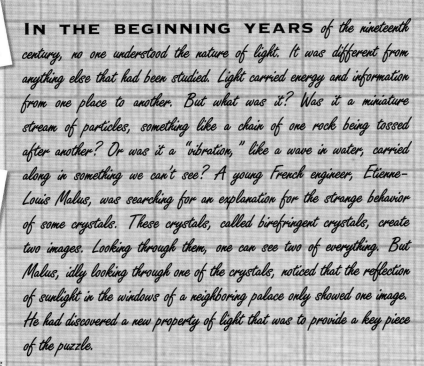

The angle at which light gets refracted depends on the shape of the material's surface.

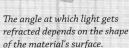

MALUS'S CAREER TAKES SHAPE

As a young man of promise, Etienne-Louis Malus attended the French military engineering school at Méziéres, where he was recognized as an exceptional student, but was unable to take advantage of a commission in the army because of political intrigues. He enlisted as a private, where, like privates from time immemorial, high on his list of duties was ditch digging. While engaged in a particular excavation, he organized the activity in such a logical, efficient manner that his superiors recommended him to become one of the first students at the recently founded École Polytechnique in Paris. He further distinguished himself there, and upon graduation enlisted in the army once again, soon finding himself granted the rank of captain.

Malus became part of Napoleon's military foray into North Africa. The British fleet trapped and burned the French fleet, leaving more than 30,000 soldiers stuck in Africa and the Near East. Napoleon decided to make lemonade with lemons; so he created the Institute of Egypt, for the pursuit of archeological, scientific, and mathematical studies. Malus became a member. Although Malus was encouraged to pursue studies of all sorts, he was not relieved of military duties.

Light being filtered through a prism.

École Polytechnique, Paris, where Malus was one of the first students.

AN ACCIDENTAL DISCOVERY

It was while under siege that Malus contracted the bubonic plague, a near certain death sentence.

Luckily for posterity it was only "near" certain, because he survived and, while recovering, composed a study of the propagation of light. The study was noted for its rigorous mathematical treatment of the movement of light. After his return to France in 1801, Malus continued, in his free time, to refine his mathematical treatise on the rules governing the movement of light beams. He published it in 1807, establishing a reputation as an expert in the field. So it was only natural he would take on a new challenge.

The "birefringent" crystals noted above are so-called because they exhibit "double refraction;" that is, they bend light not just in one direction, but in two different directions. Looking through one of these crystals, viewers see a double image. This was such an unusual phenomenon, the French Academy of Sciences was intrigued enough to offer a prize to whoever could "give a mathematical theory, confirmed by experiment," of double refraction. Malus eagerly sought the prize. He would win the prize, but along the way he was to make a more important accidental discovery.

Light refracted in a wine glass.

BIREFRINGENT CRYSTALS

He studied the transmission of light through Iceland Spar, one of the birefringent crystals. One day, as he idly observed the scene outside his window, he noticed something unusual about the reflections of the sunlight in the windows of the neighboring Luxembourg Palace on the Rue d'Enfer in Paris. When he held the crystal in a certain way, it didn't transmit two images, only one!

FACING PAGE: *Birerefringent crystals seen through a microscope with polarized light.*

Birefringent filters are used as spatial low-pass filters in electronic cameras, where the thickness of the crystal is controlled to spread the image in one direction.

Rather than simply note it as an interesting curiosity, Malus began a series of investigations into the new property he had discovered. He discovered there was a specific angle at which the reflection would provide only one image through the birefringent crystal. He discovered that the same thing would happen with a reflection off water, although the angle was slightly different. Both sunlight and artificial light acted the same way. It was as if every beam of light was made up of two parts traveling right with each other, parts that would get split apart in birefringent crystal. But when Malus found that one of the parts "disappeared" when reflected at a certain angle, he knew he had identified a universal characteristic of light, rather than a property of the crystals. Malus called this property "polarization." Years of hardship had led Malus to prepare himself to understand the propagation of light. A chance observation while pursuing a scientific prize had given him the opportunity. And Malus showed his desire by the exhaustive study he undertook. By his diligence, Malus provided a key piece to our understanding of the nature of light.

Is it a wave or is it a particle?

What did they know about light in 1800? They understood that light travels in straight lines, that white light is made up of different colors, that it travels very fast. But some things that light did made it seem as if it was a wave. For example, if there's a smooth pool of water and two pebbles get dropped in, the ripples will add and subtract, making something called an interference pattern. Light does the same thing. But there was one big problem: if light was like a wave on the water, where was the water?

That troubled Newton so much that he thought light must be made of little particles, "corpuscles." The corpuscles could act just like tiny baseballs being thrown through the air. They didn't need anything to carry them along. A pretty attractive characteristic.

Polarization was also, at first, better explained if light was corpuscular. There could be two different shapes of corpuscles: One could belong and skinny "up and down" and the other could be long and skinny "left to right."

It was a puzzle, and polarization was an important piece. The puzzle wouldn't truly be solved until new knowledge was gained in the next 100 years. And the answer would be: light was both a wave and a particle.

A nineteenth-century demonstration of refraction.

A rainbow is the result of light refracted through raindrops which act as prisms.

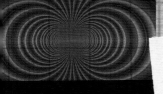

HANS CHRISTIAN OERSTED

CONNECTS ELECTRICITY
AND MAGNETISM

1820

A LECTURE ON CURRENT AFFAIRS

HANS CHRISTIAN OERSTED *was seduced by the beauty of the philosophical principle that all the forces of nature were manifestations of a single force. In particular, he believed that electricity could make heat and light and a magnetic field. That's what he was hoping to show in a lecture–demonstration in April, 1820. Regrettably for him, all his assumptions were wrong and his experiment couldn't work. Happily for him, he accidentally did an experiment he wasn't planning on, an experiment that provided the first demonstration that electricity and magnetism were two faces of the same force.*

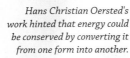

Hans Christian Oersted's work hinted that energy could be conserved by converting it from one form into another.

IN HIS FATHER'S FOOTSTEPS

Hans Christian Oersted was born in 1777 in southern Denmark, the son of a pharmacist. Hans followed in his father's footsteps, achieving academic distinction on his way to his degree in pharmacy. But he didn't stop there. University educations of the time were very broad, and Oersted got his doctorate with a philosophical thesis entitled *On the Form of an Elementary Metaphysics of External Nature*. His thesis was more or less a commentary on some ideas of Immanuel Kant, who maintained that forces of nature that appeared to us to be separate fundamentally sprang from the same source.

After graduation, Oersted received grants and scholarship money to study abroad. While in Germany, he met Johann Wilhelm Ritter, who was trying to construct physical laws that could be integrated with spiritual philosophy. Ritter also believed in a connection between electricity and magnetism. That idea stuck with Oersted.

On his return to Copenhagen, Oersted lectured. At a time when "freelance lecturers" were paid by their audience members, his talks were well-attended. It didn't take long for University of Copenhagen officials to notice his popularity and offer him a position on their staff. Oersted lectured on whatever subjects he happened to be interested in at the time, and in the spring of 1820 he was interested in the unity of electricity and magnetism.

Hans Christian Oersted, at the university of Copenhagen, discovers the deviation of a compass needle when subjected to an enclosed electric current.

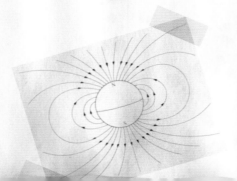

VÉRITABLE EXTRAIT DE VIANDE LIEBIG.

L'ÉLECTRICITÉ.

5. Oersted démontrant à ses amis la déviation de l'aiguille magnétique sous l'infuence d'un courant électrique (1820).

Voir l'explication au verso.

Hans Christian Oersted observes the link between electricity and magnetism, revealing a unification in electro-magnetism that led to modern machines.

ORDINARY DEVICES

Now we live in a time when all sorts of ordinary devices—from coffeemakers to automobile door locks to electric clocks—work only because of the connection between electricity and magnetism, but in 1820 there wasn't any real evidence for such a connection. Electric current could make sparks and heat wires, and magnets could pick up pieces of iron. Why should they be connected? For philosophical reasons Oersted hoped they were.

He was a believer in the "heat hypothesis." When current ran through a wire, the wire converted the electricity to heat. A thin platinum wire could get so hot it would glow. Oersted believed that a wire throwing out heat and light would also throw out energy in the form of magnetism. He decided that a compass needle, which aligns itself with a magnetic field (usually the magnetic field of the Earth), would be a good detector. So he planned to heat up a wire and put a compass needle nearby and see if the needle tilted.

Even common household devices, like this coffeemaker, have switches and valves that work only because electrical energy can be converted into magnetic energy.

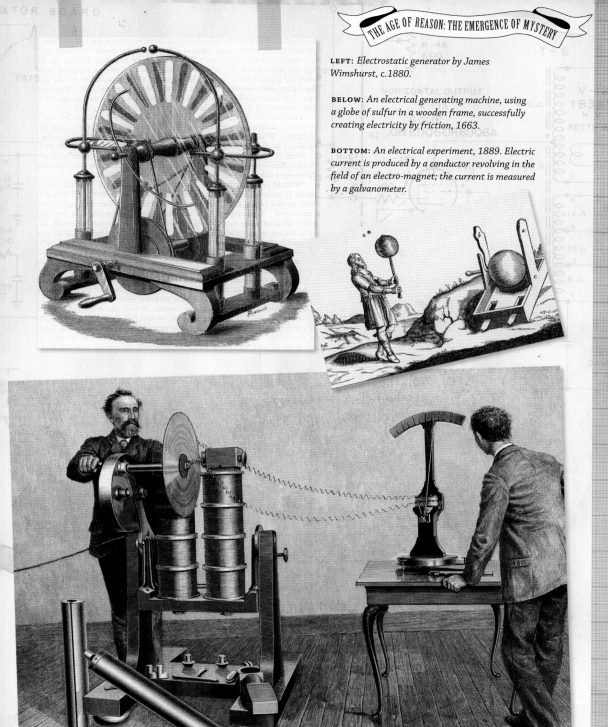

LEFT: *Electrostatic generator by James Wimshurst, c.1880.*

BELOW: *An electrical generating machine, using a globe of sulfur in a wooden frame, successfully creating electricity by friction, 1663.*

BOTTOM: *An electrical experiment, 1889. Electric current is produced by a conductor revolving in the field of an electro-magnet; the current is measured by a galvanometer.*

Because there must be a connection between heat, electricity, and magnetism, he figured this would be a great demonstration for the people who came to his lecture. There was only one problem. Either because of a lack of time or a lack of resources, he didn't test his demonstration ahead of time.

SHOCKING CONCLUSIONS

There was Oersted, standing in front of his class, lecturing on the connection between fundamental forces. No one knows exactly what demonstration he was trying to do, but it wasn't the one he ended up with. He hooked a platinum wire up to a battery, waited for the wire to glow white hot, and then moved the compass into position. While he was moving the needle into position he accidentally moved it right above the wire where it turned sideways. The effect was small, the compass was in the wrong spot, and it moved the wrong way. But Oersted noticed it. This was the first evidence that magnetism could be created by electricity.

Oersted has been criticized because he waited to do his next experiments until July—after all, he had just connected two different forces into one!—but the effect had been so small he needed to assemble larger batteries to study it properly. He was also hampered because he expected a result that would connect heat with magnetism, and he was looking for the wire to "push out" a magnetic field the same way it pushed out heat. In a new series of experiments he didn't find either of these results. To his credit, he didn't allow his preconceptions to blind him to what he was seeing. In a groundbreaking paper he published a couple of shocking conclusions. First, any wire carrying current generated a magnetic field, and second, unlike any force yet discovered, the magnetic force lines were circular, going around the wire.

Oersted was continually inspired by his philosophical beliefs, but when opportunity came to him, he didn't limit himself to investigating just what he believed to be true. His desire pushed him to follow his observations into a new realm. His discovery, linking electricity and magnetism, is the fundamental principle enabling everything from microphones to electrical power generators.

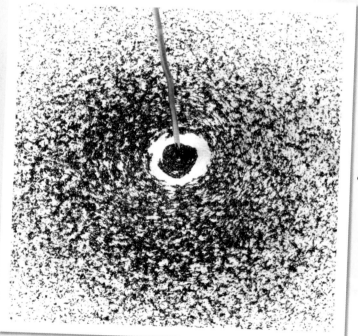

Hans Christian Oersted discovered that magnetic force lines around a current-carrying wire are circular.

Moving or not?

Oersted's discovery was remarkable and essentially unexpected, and his investigations were convincing enough to generate an entire new branch of scientific and technological development. But he failed to develop a mathematical framework for his discovery. The mathematical formalism had to wait for the French scientist André-Marie Ampère, who provided a theoretical explanation for the generation of a magnetic field. According to Ampère, any moving electric charge created a magnetic field. But, again according to Ampère, magnetic fields were only created by moving electric charge, that is, only by electric current. So how to explain permanent magnets?

Well, there was no explanation. Not for about 100 years. That's how long it took to connect up knowledge about the form of atoms with our observations of magnetic phenomena. Now we know that every atom has a "stationary" nucleus with electrons whizzing around. The "current" created by those whizzing electrons creates a magnetic field, but in most materials those fields get canceled out. In a few materials, such as iron, the fields can line up with each other, creating a "permanent" magnet. But, as would no doubt satisfy Oersted's desire to unify nature's forces, at the basic level even permanent magnets are the result of electrical currents.

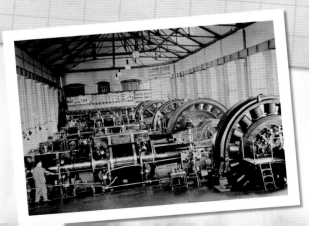

A view of a Turbine hall at the Barcelona Electricity Company, 1897.

CHEMISTRY

BIOLOGY

FRIEDRICH WÖHLER

SYNTHESIZES ORGANIC
MOLECULES

1828

Organic compounds are
made up of molecules
containing carbon.

BRINGING THE
LAB TO LIFE

★

IN ANCIENT TIMES THERE WAS A BELIEF
*that living beings were distinct from non-living materials because
of the presence of an undefinable "vital" force. Vitalism held that
beings were alive because of this vital force, meanwhile they showed
they had vital force because they were alive. This circular reasoning
bothered scientists of the Age of Enlightenment, but the alternative was
a kind of mechanical explanation that had problems of its own.*

Liebig at work in his laboratory at Giessen,
Germany, which he established as the first
practical chemical teaching laboratory, 1840.

MECHANISTIC THEORY

The mechanistic theory held that living beings were like machines, containing miniature devices that would convert food to energy, contract and expand muscles, and even transmit information from one part of the body to another. Now we know the mechanistic theory is true, but in the early 1800s none of the "miniature devices" within living beings could be identified; so there wasn't a whole lot of evidence against vitalism. And there was one big piece of evidence for vitalism: living beings could produce chemical compounds that couldn't be made in test tubes and flasks. These organic compounds were signs that chemistry inside the body worked by different rules than chemistry outside the body. The difference was defined by the presence of the vital force.

This was the worldview that Friedrich Wöhler was going to shatter.

Wöhler was born in 1800 near Frankfurt. He studied medicine in Heidelberg, then went to work under one of the foremost chemists of the day, Jöns Jakob Berzelius of Stockholm. Wöhler, like all educated chemists of the day, knew the principles of vitalism, the doctrine of vital force. He had seen and studied a variety of organic compounds, and an equal variety of inorganic compounds.

Jons Jakob Berzelius (1779–1848), scientist and chemist.

An engraving of Frankfurt's main square, c.1840.

Heidelberg University students, members of the Allemana society.

In 1828 Wöhler was trying to make a big batch of an inorganic compound known as ammonium cyanate so that he could learn about its possible uses. He was able to make ammonium cyanate by mixing a couple of different inorganic compounds, such as lead cyanate and caustic ammonia, or silver cyanate with ammonium chloride. When he looked at the product of the reaction, he found a crystalline form that was familiar. It was urea.

The human body breaks down proteins into amino acids, and continues to break down amino acids into smaller molecules. The final product is ammonia, a toxic compound. Because ammonia is toxic, the body has to get rid of it; so the liver takes the next step by chemically modifying ammonia and turning it into urea. The urea is cleaned out of the body and excreted in the urine. In the early 1800s, nothing was known of amino acids or the chemical functions of the liver, but urea had been identified as an extract of urine, in humans and in animals. And nowhere else. Urea did not go into humans, but it came out. So it had to be made within a living organism, which means it was—by

definition—an organic compound. But Wöhler had made it—not by plan, but by accident—from inorganic compounds. He summarized his discovery in an excited letter during February, 1828, writing to his friend and mentor, Berzelius:

I cannot, so to say, hold my chemical water and must tell you that I can make urea without the help of a kidney or even an animal, neither man nor dog. Ammonium cyanate is urea.

From our modern perspective we can perhaps be forgiven for not getting excited about a chemist being able to make something that all of us make every day without the help of a test tube or a laboratory flask. But this mundane, simple, and relatively unimportant chemical synthesis was the first sign that chemistry within the body and chemistry outside the body were governed by the same rules. It wasn't necessary to postulate the existence of an entirely separate force, a separate set of chemical rules.

Wöhler's training had prepared him to recognize the significance of the chemical reaction that chance placed in front of him. His discovery was the first step toward developing an understanding of the biochemical processes that allow us to live our lives.

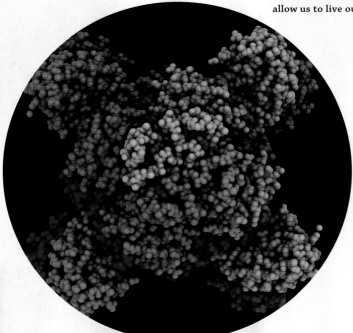

Friedrich Wöhler's discovery was a landmark in the history of science which disproved and undermined the Vital Force Theory which had been believed for centuries.

Killing vitalism

"Vitalism," the belief that a being is alive only because of an unmeasurable "vital force," did not lose all its adherents with Wöhler's successful inorganic synthesis of an organic compound. Vitalism holds that there must be some driving force within living organisms other than their biochemical reactions. It's an attractive idea, but it's not a scientific idea. For a theory to be considered scientific, it must be falsifiable. That is, it must be possible to perform a test that would fail if the theory is incorrect. The proof of vitalism goes something like this:

"That creature is alive, therefore it has vital force within it."

"How can you tell it has vital force within it?"

"It's alive."

YELLOW PENCIL

There's no test to disprove that kind of circular logic; so vitalism is not a scientific theory. Still (for various reasons), it remained around for about 100 years, slowly being erased as each organ's function was understood in terms of biochemical reactions. Perhaps appropriately, the last organ whose function was "unexplained" was the kidney, an organ involved in the production of urine. In the 1930s, the filtration and secretion functions of the kidney were demonstrated to be biochemical, just like the functions of the rest of the body. Vitalism was dead and the body was understood to be a biochemical organism.

CHARLES GOODYEAR

INVENTS THE RUBBER
VULCANIZING PROCESS

1839

FIXING A STICKY SITUATION

Throughout tropical regions, farmers slice rubber trees to harvest the latex within.

IN THE EARLY 1800s *a rubber craze swept Europe and the United States. This unique material was pliable, waterproof, and elastic. But in cold temperatures it would become brittle, and in hot temperatures it would melt. Charles Goodyear was among hundreds of people who tried for years to fix these problems, but his perseverance led to a happy accident which allowed him alone to succeed.*

Collecting latex in Para, Brazil. The tree is tapped and the sap collected.

RUBBER FACTORIES

Latex was brought to Europe from South America. People in South America had been making shoes, capes, and even bouncy balls with the elastic material they produced from the sap of the *Hevea brasiliensis* tree. The latex material dried quickly, though, and became difficult to work into other forms. Kneading it and dissolving it in turpentine created a liquid that could be worked. In 1770 Joseph Priestley, an English chemist, noticed that pieces of the dried latex were useful for "rubbing out" pencil marks. The name "rubber" stuck to the new material. Unfortunately, on a hot day, other things stuck to the new material as well.

Still, it was exciting enough that hundreds of "factories" sprang up. Walking the streets of New York in 1834, 34-year-old Charles Goodyear happened across a store selling life preservers. Goodyear saw their poorly designed valves and returned to the store a few weeks later with an improved design. The storekeeper had no interest and showed Goodyear why: a heat wave had come, melting the life preservers into a sodden, stinky mass. Rubber was useless. That pile of sticky, stinking rubber was to change the world. Charles left the store with a new mission: to find a way to convert rubber into a material that would stay pliable, waterproof, and elastic in cold winters and hot summers. He also left that store the way he had come in, with debt hanging over his head. A creditor had him arrested and tossed into debtor's prison. He took with him some pieces of rubber and his wife's rolling pin, to knead the rubber.

For the next few years Goodyear struggled to develop a process for treating rubber. He flirted with brief periods of limited prosperity, but spent most of those years (indeed, most of his life) hovering between poverty and destitution. The constant was his dedication to fixing the rubber problem.

First he found that mixing magnesia into the rubber produced a smooth surface with a pleasing white color. An investor fronted him money, then he and his wife made hundreds of rubber-coated shoes. Then he decided it might be a good idea to determine if they would last. When spring turned to summer they melted to a sticky mess. He lost the money and his reputation.

He continued to mix materials with rubber, and he also tried to make his rubber objects aesthetically pleasing. He painted some with bronze paint. He decided to re-use a painted piece, etching away the paint with nitric acid. The acid-etched piece looked worse; so he threw it away. Later he ran across it in the trash and noticed the surface was smooth and hard. He convinced another investor to come up with cash. Goodyear turned out hats, aprons, life preservers...all of which turned sticky and gooey once high summer hit.

THE BREAKTHROUGH

Goodyear stuck with it. He worked with Nathaniel Hayward, a Massachusetts stable owner who had become intrigued by rubber's properties. Hayward found that by sprinkling sulfur on a rubber sheet and letting it cure in the sun, the surface would become smooth and hard. With new investors and the new process they sold goods, including 150 rubber mailbags for the United States Postal Service—bags which drooped and sagged at high temperatures. The surface still retained its shape and smoothness, but the interior just melted away.

The failure drove the last of his investors away, but it didn't drive away Goodyear's desire. If sulfur on the outside made a good surface, wouldn't it work on the inside as well? He mixed and kneaded different proportions of sulfur and rubber. Nothing worked.

There are several versions of the story of Goodyear's accident, the chance that led him to great success.

Possibly Goodyear was angered at being teased by a group of townsfolk and flung his arms, sending rubber he'd been kneading flying onto a stove. Another story has it that a pile of returned goods got tossed near a stove, where some of the rubber-coated material "cooked." Another version simply has Goodyear spilling part of the mixture onto a hot stove. But, whatever the story, the result is this: some of the sulfur-impregnated material accidentally ended up on a stove where it cured. Cured. Not melted. He immediately recognized the significance. Heat converted the sulfur-rubber mixture into a durable, stable material retaining all the useful properties of rubber.

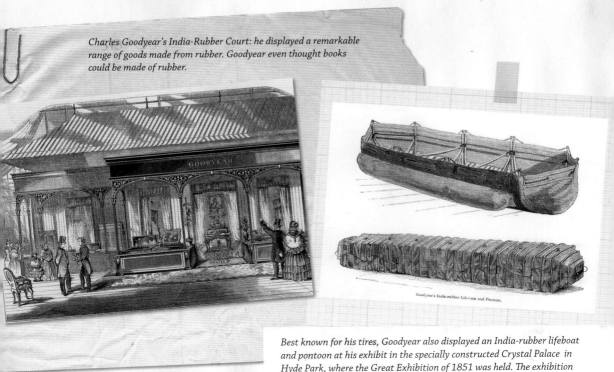

Charles Goodyear's India-Rubber Court: he displayed a remarkable range of goods made from rubber. Goodyear even thought books could be made of rubber.

Goodyear's India-rubber Life-boat and Pontoon.

Best known for his tires, Goodyear also displayed an India-rubber lifeboat and pontoon at his exhibit in the specially constructed Crystal Palace in Hyde Park, where the Great Exhibition of 1851 was held. The exhibition was conceived by Prince Albert, in order to showcase the industrial, military, and economic superiority of Great Britain.

DETERMINATION

Because his investigations had given him the preparation, Charles Goodyear recognized the significance; but everyone else was tired of hearing about it. It took him several months, maybe even years, to identify the right mixture and processing temperatures to make what was eventually called "vulcanized" rubber. His desire had never flagged; so when the opportunity presented itself he was ready to take advantage. Rubber is now used in just about every manufacturing facility in the world, in all sorts of apparel, and, of course, in tires. All because of a spill on a stove.

Slick solutions

Sometimes inventions or discoveries require so many preliminary steps that it seems astounding they ever became possible. For example, Charles Macintosh, a Scottish chemist, invented the raincoat that still bears his name, albeit in slightly different form. Macintosh was manufacturing ammonia for industrial use, starting from wagonloads of human urine. He (and his workers!) were eager to find another raw material to work from. At that same time, coal gas became the lighting solution of choice in England, producing tons of nasty stuff called coal tar.

Ammonia could be synthesized from coal tar; so Macintosh switched from urine to coal tar. But after taking the ammonia out, he was left with naphtha, a volatile solvent. Wanting to find some use for tons of naphtha, Macintosh found it could dissolve India Rubber into a liquid, which he could then use to coat cloth to make a waterproof surface. Waterproof, but sticky. So Macintosh added a top layer of cloth to make a rubberized cloth sandwich, which ended up as the waterproof raincoat known as the mackintosh.

IGNAZ SEMMELWEIS

REDUCES THE INCIDENCE
OF CHILDBED FEVER

1847

THE FIRST "MISTER CLEAN" IS NOT SO WELCOME

IN 1847 THE SOUND OF BELLS *almost literally drove Ignaz Semmelweis crazy. He was in charge of the first delivery ward at Vienna's Maternity Hospital, a ward in which more than one out of every ten women entering the hospital would die. The sound of the priest's bells administering last rites drove Semmelweis to believe "life seemed worthless." To rejuvenate his spirits he took a trip to Venice, but while there a colleague of his died back in Vienna. That accidental and tragic death set Semmelweis on the path to save hundreds of lives. If the medical establishment had been ready to accept his results, many thousands more could have been saved.*

A sixteenth-century birthing stool, on which the mother sat while her child was delivered.

CHILDBED FEVER

Ignaz Semmelweis became a doctor in 1844 and began working at Vienna's free Maternity Hospital. The hospital had two delivery wards. The first was attended by medical students and their respected teachers. The second ward was attended by midwives. Yet while one out of ten mothers delivering in the first ward would end up dead, only one out of 50 or 100 of those delivering in the second ward would be struck down. The disease, puerperal fever, also called childbed fever, struck following or sometimes even before delivery. Mothers, and sometimes their newborn children, suffered from inflamed lymphatic and blood vessels and additional inflammation around the lungs, heart, and abdomen. Death soon followed. Because no one knew what caused the disease, no one knew how to stop it. Theories abounded.

Perhaps it was "miasma," or bad air, that created conditions leading to disease. Or maybe the etiology—the source and progress—of the disease was due to melancholy temperament, or overcrowding, or the wrong birth position, or even constipation. But the list doesn't stop there. Other explanations: mothers' embarrassment at having their genitals seen by the male doctors, fear because of the reputation of the first ward, maybe even the act of conception itself induced unknown changes in the blood. It's no wonder nothing could be done to stop the disease, because each patient's death could have been due to any of dozens of causes. Each new death stimulated

a new explanation. When mother and newborn both perished, each of those deaths had a separate explanation as well. Semmelweis was put in charge of the ward in 1846, understanding only that nothing was understood. After seeing hundreds of victims of childbed fever treated unsuccessfully, Semmelweis concluded that the explanations bandied about "cannot contain the actual causal factor of the disease." But he was no closer to an explanation. So he suffered with each new death, feeling each as a failure, a stain on his reputation, and a stain on the reputation of the first ward.

Infant mortality was both feared and expected, so one strategy for making sure you had children to outlive you was to have a large family, in the hope that some would survive to adulthood.

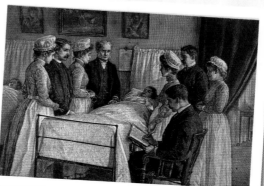

Ignaz Semmelweis's observations conflicted with the established scientific and medical opinions of the time.

ABOVE: *Incubators at the Maternity Hospital, Port Royal, Paris, in 1884.*
RIGHT: *A US postage stamp issued in 1999 commemorating the postwar baby boom of the 1940s, '50s, and '60s.*

CADAVEROUS PARTICLES

The reputation did not need any more staining. The hospital admitted patients to either the first or second ward on alternate days of the week. Expectant mothers knew the schedule and the reputation of the first ward; so many chose to deliver their children in the street, claiming they could not make it to the hospital in time. They then came in for post-natal care—and they didn't suffer from childbed fever! Nothing could explain this.

So when Semmelweis had the opportunity for a brief sabbatical, he headed to Venice, hoping the beauty of the art would revitalize his soul. But while he was gone his colleague and associate, Jakob Kolletschka, had an accident and died. Kolletschka was a professor of forensic medicine. He'd been conducting an autopsy when a student inadvertently cut Kolletschka's finger with a scalpel. Soon fever hit, and he was dead. The autopsy of Kolletschka's body showed...inflamed lymphatic and blood vessels and additional inflammation around the lungs, heart, and abdomen, along with other symptoms identical to puerperal fever.

It suddenly became clear: there wasn't a separate cause for each mother's and each newborn's death; they all died from the same cause. The same cause that had killed Kolletschka. The circumstances of his death provided the final clue. Some "cadaverous particles" had entered into his bloodstream.

Now everything began to come together. The physicians and students would go back and forth between the first ward and the autopsy room, where they examined corpses—some of which had died from childbed fever. Even if they washed, their hands would still smell of cadaverous material; so Semmelweis reasoned they were carrying cadaverous particles from patient to patient—the doctors killing the very patients they hoped to save. He instituted extra cleaning procedures, essentially having the doctors clean their hands with what we now know of as household bleach.

Death rates plummeted. The first ward had months with no deaths at all. Long-term death rates from childbed fever evened out between the two wards. Semmelweis was a hero. Or he should have been. But the medical establishment could not accept Semmelweis's radical interpretation of the cause of the disease. And even worse: he blamed the deaths on physicians! Careful observation had prepared Semmelweis to properly interpret the accidental opportunity presented to him; and his desire to fix the problem was palpable. But his desire was not shared by his colleagues and superiors; they hounded him from his position, repealed his reforms. Death rates rose. If he had been able to offer something other than the vague "cadaverous particles" explanation he might have prevailed, but the germ theory of disease was still to come. After his death, Joseph Lister, who helped prove bacteria caused disease, said, "without Semmelweis, my achievements would be nothing."

Although Semmelweis was not the first to broach the idea of a contaminant carried by physicians, he developed what appears today to be an unassailable argument in support of his explanation. He examined detailed statistical records and correlated them with changing practices. He investigated mortality rates for both mothers and newborns. And he documented the changes in mortality rates with the new procedures he implemented. But without an underlying explanation of how the disease was carried, his case did not carry the day.

Ignaz Semmelweis died at the age of 47, in an asylum, perhaps driven there by his frustration. One prominent story holds that he died from an infection very similar to the puerperal fever he had fought against.

Sir Thomas Watson

Proving his case

Semmelweis was not the first to suspect some kind of contaminating agent was being transmitted by caretakers. Thomas Watson, an English medical professor noted that "diligent ablution" should be practiced by medical attendants when puerperal fever was present. Oliver Wendell Holmes, the Boston doctor, claimed puerperal fever was transmitted from patient to patient by their attending physician. He was severely rebuked by the medical establishment, being told that any self-respecting doctor was a gentleman and, "a gentleman's hands are clean."

WILLIAM HENRY PERKIN

INVENTS THE FIRST
ARTIFICIAL DYE

1856

A NEW WORLD STARTS WITH A DYE

ABOVE: *Powdered dyes on sale in a market.*
BELOW: *Evening wear from the 1860s, just a few years after the invention of the first synthetic dye.*

IN THE 1850S THE INDUSTRIAL REVOLUTION was in full swing in Britain. Coal-fired steam engines powered manufacturing in many industries. The textile industry, in particular, had seen explosive growth. Enough raw material was available that the idea of a "wardrobe"— multiple garments of the same type—was spreading beyond the upper classes. Although the cloth was available, the capacity to produce dyes for coloring all that cloth had not kept pace. The industry was crying out for new ways to make their products more attractive. At the same time, Britain had another problem.

WILL'S CIGARETTES.

TINCTURE OF QUININE

THE BARK AND FLOWERS OF CINCHONA.

Quinine (Cinchona): an invaluable medicine in tropical countries; excellent for reducing fevers and used to control malaria.

SYNTHETIC DYES

The stretch of the British Empire reached into tropical lands, exotic tropical lands with exotic tropical diseases, including malaria. The only treatment involved quinine, an expensive, hard to refine compound. While trying to develop an inexpensive method for producing quinine, the teenager William Henry Perkin instead accidentally invented the first synthetic dye. The accident seeded all the chemical industries that were to follow.

As a 15-year-old, William Henry Perkin was excited about the potential of chemistry. In 1853 it was fairly well understood that all the materials in the world were composed of the same building blocks, the chemical elements, but very little was known about how those building blocks went together. Even though his father, a builder, thought chemistry was a waste of time, Perkin began studying under August Wilhelm Hofmann at the Royal College of Chemistry in London. Perkin was excited about chemical analysis: developing tests to split compounds into their separate elements. He was equally excited by chemical synthesis—generating new ways to bring separate molecules together to make new ones. He was so excited he made his own laboratory in an upstairs room in his father's house.

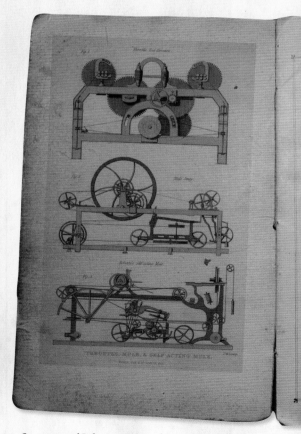

Crompton and Roberts cotton machinery, 1835.

Mine workers ready to start work in a coal mine, having just arrived at the bottom of the shaft, 1855.

Edmund's, the main colliery in Yorkshire.

In the 1850s, Britain was coal-driven. Much of the coal was converted to coke, a hard, hot-burning extract of raw coal. Processed coal also yielded "town gas," which was used for artificial lighting. Coke production also left behind a stinky, thick, noxious waste product: coal tar. Perkin's teacher, Hofmann, was an expert on coal tar. He knew about the compounds that could be easily created from coal tar, and he knew which elements composed those compounds. The compounds in coal tar were composed mostly of carbon and hydrogen (which is why coal and similar materials are known as hydrocarbon fuels). Hofmann knew that quinine was also mostly carbon and hydrogen, with a little nitrogen and oxygen thrown in; so he figured it might be possible to make quinine from coal tar. So, when Hofmann went to Germany for spring break in 1856 his 18-year-old student, Perkin, figured he'd surprise and impress his teacher. He brought some coal tar to his home laboratory and tried to make quinine. He failed.

Perkin started with naphthylamine, a coal tar compound that had about quinine's ratio of carbon to hydrogen to nitrogen, it just needed a little more oxygen. He tried different ways of treating the coal tar sludge, but all he came up with was something like more coal tar sludge. So he moved on to another coal tar compound, aniline. Again he tried ways of adding more oxygen, and again he ended up with a black sludge. No one today knows why he did what he did next. Maybe he was just playing. Maybe he was desperate to prove to his father and his professor that he wasn't wasting his time. Maybe he was just desperate to show something for all the work he'd put in. He took that black sludge and mixed in a little ethanol. The resulting liquid had a rich purple color. He dipped cloth into the purple liquid and the color was taken up. He sent a sample of his dye to a silk manufacturer. The manufacturer wrote back that the color was beautiful, and that if it resisted fading as well as it appeared to then Perkin would be a rich man.

The Industrial Revolution saw great change and invention in the textile industry around the world—from the invention of dye and industrial spinning machines in the UK to the cotton gin, as shown above, by the US inventor, Eli Whitney Jr.

The manufacture of aniline red dye.

143. DYE-BECKS.

Dyeing cotton fabrics in dye-becks.

MAUVEINE

Perkin dropped the search for quinine and worked with his brother and a friend on methods of chemically synthesizing his new purple dye, which he called mauveine. Within a few months the three young men convinced themselves they could produce mauveine inexpensively. Better yet, Perkin convinced his own father that chemistry was a valuable and potentially lucrative occupation.

Perkin had earned his father's approval, but, more important, he had also earned his father's investment. The senior Perkin funded a chemical factory for his son. William Perkin quit his work at the College, earning himself the scorn of his academic mentor, Hofmann. In a few short years Perkin was to prove he had made the right decision. His new dye brought rich color into the lives of millions of people, and brought a fair amount of richness into Perkin's life as well. In the face of discouragement, his desire allowed him to recognize and exploit his chance discovery.

Nineteenth-century illustration of the process of dyeing. Published in Specimens des divers caractères et vignettes typographiques de la fonderie by Laurent de Berny (Paris, 1878).

Cashing in on chemistry

At the time of Perkin's discovery of mauveine, chemistry was, as a famous chemist was to claim, not a science, but "widespread dilettantism," especially in England. After Perkin established his mauveine production, chemistry became a sound foundation for business. The textile industry was crying for large quantities of high-quality dye, and Perkin answered their need. He had taken chemistry out of the lab and brought it to the factory floor.

Although chance played a large role in his initial discovery, Perkin did not relax. He developed several other dyes in the following years. His commercial success encouraged others to develop industrial-scale chemical manufacturing for everything from soap to explosives. Perkin himself had additional success with new dyes, so much success that he was able to "retire" when he was 36 years old. He didn't rest, however. His fortune made, he returned to the chemistry research lab. Although his direct influence on the world was great, his indirect influence is even more impressive. His example proved that chemistry was an economically valuable pursuit, and companies around the world were formed to identify and commercialize chemical synthesis.

RUB-A-DUB-DUB, THREE MAIDS AT A TUB,
FILLED WITH DYE MADE FROM MAYPOLE SOAP,
THEY'VE DIPPED ALL THEIR CLOTHES AS EVERYONE KNOWS,
GREEN, YELLOW, AND PALE HELIOTROPE.

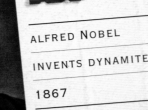

ALFRED NOBEL

INVENTS DYNAMITE

1867

SIDESTEPPING FATE

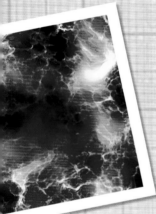

NITROGLYCERINE. *More than 100 years after it has been in common use, the word still conjures images of danger, of liquid flasks so sensitive that a slight nudge can bring explosive disaster. So perhaps we can be forgiven for believing that, when a full flask slipped out of his hands to fall on the floor, Alfred Nobel saw his life flash before his eyes. Although that story has circulated, that's probably not exactly what happened, but another near-death experience put Nobel on the path toward inventing what he thought would be his most humanitarian invention: dynamite.*

THE INITIATOR

Alfred and his younger brother Emil followed in their father's footsteps, taking up careers developing explosives. In 1862 Alfred developed a process that allowed the large-scale manufacturing of what he called "blasting oil"—glycerol trinitrate, or nitroglycerine. Nitroglycerine did not reliably detonate with the kinds of fuses that would trigger gunpowder; so Alfred invented a reliable method for triggering nitroglycerine: the "initiator," later to become the "blasting cap." These inventions came at a time when large scale rail and port construction and mining were ripe for growth, and Nobel's innovations found a ready market. Things were going well.

EXPLOSIVE TRAGEDY

Until September 3, 1864. That's the date Nobel's factory in Heleneborg, outside Stockholm, exploded. Alfred escaped injury, but five others were killed, including his younger brother Emil. Alfred continued to believe in the value of nitroglycerine, but even as he was on an 1866 tour of the United States to demonstrate the safety of properly handled blasting oil, he received

Explosives are used today by the construction and mining industries.

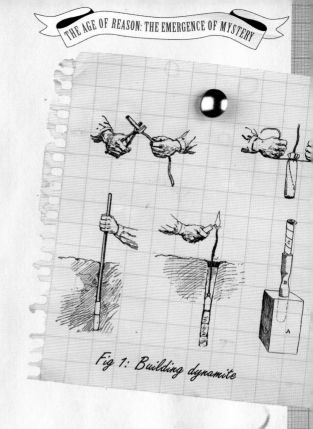

Fig 1: Building dynamite

Putting dynamite into cartridges at a
dynamite factory.

word that there had been an explosion at his
factory in Krümmel, Germany. Yes, he reasoned,
nitroglycerine is safe when handled properly,
but if we can't always handle it properly even in
our own factories, how can we expect others to
be more conscientious?

A CHANCE OBSERVATION

If only there were some way of improving the
stability of blasting oil! The exact details of what
came next may never be known, but a chance
observation by Nobel seems to be critical. A
standard shipping method for flasks of blasting
oil was to crate them in sawdust. From time to
time a container would leak and the oil would
be absorbed by the sawdust. Nobel, noticing
that nitroglycerine mixed with sawdust did not

appear to suffer from instability, checked that it
would still detonate with a blasting cap. When
he found that it did, he initiated an intense
development program to identify the optimum
neutral absorbent material that would mix with
nitroglycerine to be both powerful and safe.

He found, again by chance, that the particular
type of soil outside the Krümmel factory was
just about perfect. The soil, called kieselguhr, is
made up of diatoms, tiny algae whose skeletons
are made of silica, just like sand on the beach.
But unlike sand the kieselguhr is very absorbent.
When mixed with blasting oil, it creates a thick
putty that can be safely handled, but retains
most of its explosive power. Dynamite was born,
and with it, Nobel's fortune.

Manufacture of dynamite at Nobel works, Ardeer, Scotland.

LEGACY THAT
ALMOST WASN'T

Alfred Nobel's inventions came at the perfect time to support industrial expansion. The world was hungry for a safe, efficient way of rapidly building tunnels, recharging oil wells, dredging canals, and more. Nobel fed that hunger, and was rewarded with a fortune. The explosives he developed were used not only for peaceful purposes, but in war as well. Nobel was quite aware of the wartime uses of his explosives, and was active in the development of munitions. But he thought his work would bring the world closer to peace:

Perhaps my factories will put an end to war...
on the day that two army corps can mutually
annihilate each other in a second, all civilized
nations will surely recoil with horror and
disband their troops.

The world did not share his belief in this promise. In 1888 a French newspaper erroneously reported his death with an obituary under the headline of "The Merchant of Death is Dead." It is widely believed that seeing firsthand the legacy for which he was going to be remembered caused Nobel to do what he could to change it. He modified his will to create a foundation for administering Nobel Prizes in Physics, Chemistry, Physiology or Medicine, and Peace. Perhaps these prizes themselves are the result of a chance occurrence: an obituary published just a bit prematurely.

LOUIS PASTEUR

DEVELOPS CHICKEN
CHOLERA IMMUNIZATION

1879

CHICKEN SOUP GONE BAD—GOOD NEWS FOR HEALTH

IN APRIL OF 1878, *Louis Pasteur had lectured the members of the French Academy of Medicine, telling them that diseases were caused by infectious microbes. To his skeptical audience he described the evidence that supported his theory. Even as he was marshaling his arguments, he realized the real work lay ahead: trying to prevent disease. He was fated to make a key discovery, but the solution ended up being triggered not by hard work, but by a summer vacation.*

TOP: *French postage stamp to commemorate the hundredth anniversary of the rabies vaccine. Doctor, patient, and inventor Louis Pasteur in background.*
ABOVE: *A Polish postage stamp with an illustrated portrait of Louis Pasteur.*

FERMENTATION

Louis Pasteur was a chemist. He had begun his trip to fame by discovering that a particular chemical, tartaric acid, existed in two chemically identical forms, with one form more or less the mirror image of the other. Then he found that tartaric acid is produced in wine, but only in one of its two forms. Why, he wondered? He proposed that the "left-handed" form in wine was produced by living organisms. That led him to wonder what these living organisms were doing in the wine anyway.

Pasteur had already achieved a measure of fame from this work, and he'd been appointed Dean of the Faculty of Sciences in Lille, France. In his inaugural address to the faculty, he said what he was to say many more times in his life: "Where observation is concerned, chance favors only the prepared mind." As we shall see, these were prophetic words.

In the 1850s, the French wine industry was suffering from poor production yields. Some batches of crushed grapes fermented into wine, others fermented into lactic acid. Industry leaders sought help from Pasteur. He resumed his investigation. He looked at micro-organisms in the fermenting juice. He noticed that the batches of good wine had lots of healthy yeast and the batches of bad wine had yeast and other rod-like microbes. He also noticed that the intermediate alcohols and acids produced during fermentation were "left-handed." The conclusion was clear. It wasn't coincidence that the microbes lived in fermenting juice, the microbes themselves were causing the fermentation! This was a completely new idea, and Pasteur needed to prove it.

Louis Pasteur (1822–1895), chemist and microbiologist, in the hot room for the cultivation of microbes at the Institut Pasteur, Paris.

Louis Pasteur checking animals in his laboratory at the École Normale, Paris.

CHICKEN CHOLERA

In an elegant series of experiments, he demonstrated that boiled solutions kept isolated from microbial contamination would not ferment. Those same solutions, once yeast was added, would ferment into alcohol. And those same solutions, when open to contamination by microbe-laden dust particles, would spoil.

After years of investigation into the problems of fermentation, Pasteur concluded that different micro-organisms did different things. Some made good beer, some made good wine, and some spoiled both. Pasteur had made a leap to identify left-handed chemicals with life. He had made a leap to conclude that fermentation was a product of bacterial action. Now he made another leap. If microbes could act externally on wine and beer (and such things as spoiled meat), then why couldn't they act internally on other living organisms?

Specifically, in a controversial 1878 address to the Academy of Medicine, Pasteur declared that most transmissible, contagious, infectious diseases were caused by microscopic organisms.

Several of these organisms had already been identified, but they were generally thought to be a result of a disease, not its cause. Many members of the Academy remained resistant to Pasteur's idea.

Meanwhile, Pasteur and his colleagues Chamberland and Roux were studying the bacterium they believed caused chicken cholera, a fatal, highly infectious disease that was devastating poultry flocks. They cultured the bacteria in a special broth, then injected chickens with an extract of the broth. The injected birds invariably died within 48 hours. It was clear to Pasteur that there was no other explanation: the microbe must be to blame for the disease. But how to cure or, better yet, prevent the disease?

FACING PAGE ABOVE: *Louis Pasteur in his laboratory.*
FACING PAGE BELOW: *L'Institut Pasteur, Paris, where Pasteur was Director between 1888 and 1895.*

PARIS. — L'Institut Pasteur

268

ND. Phot.

FINDING A CURE

In test after test, Pasteur's group made no progress toward treating the disease. Every chicken injected, died. Finally Pasteur had had enough, he needed a vacation. He told Chamberland to take care of injecting more chickens with the next batch of bacteria. Chamberland, like many assistants before and after, either forgot, or decided that if the boss wanted it done he could take care of it himself. Chamberland left on his own summer vacation. When they both returned after a few hot summer weeks, they injected a batch of chickens with the bacteria-infested broth that had been left for weeks in the summer heat. None of the chickens died.

They were a bit dismayed at the failure. All the work necessary to isolate and culture the bacteria was wasted because, obviously, the batch had spoiled. They made a new batch with enough broth to inject many more chickens so they could make up for lost time. Again, the chickens died. Wait! Some chickens died. Chickens that had earlier received a dose from the bad batch got sick, but recovered. Every one of them.

Another leap. Pasteur reasoned the "bad batch" must have consisted of weakened or "attenuated" bacteria, damaged by their extra time in the summer heat. Chickens injected with the attenuated bacteria were then protected against the full-strength bacteria. Pasteur perfected his attenuation technique and found that he could protect nearly all the animals that were injected with the weakened bacteria. Opportunity had come to Pasteur, an opportunity he was prepared to understand, an opportunity he wanted to exploit. Pasteur had developed the first intentionally-created "vaccine," as he named the new treatment, and as we continue to call it today.

THE VACCINE

A chance mistake, a couple of summer vacations, and Pasteur had discovered the principle that continues to save millions of lives.

Pasteur had developed his cholera vaccine based on his belief that vaccines worked when attenuated bacteria "used up" something within the body; something that wasn't there for the fully functional bacteria to eat when they infected the body. So he concentrated on injecting weakened live bacteria as a vaccine. Meanwhile another prominent veterinarian was having success injecting dead bacteria as a vaccine. (Now we know vaccines work by preparing the body's natural defenses to recognize molecules associated with infectious agents.) Because this didn't fit with Pasteur's ideas, he didn't believe it would work.

Pasteur was not shy about trumpeting his own success. Probably one of the reasons his name is better known than the names of many of his equally significant contemporaries is because he was his best press agent. So he made sure to announce the success of the cholera-preventive vaccine. But he encountered opposition, and plenty of it. Some was from scientists who were still resistant to the germ theory of disease. Some was from doctors and veterinarians, who thought a mere chemist could have nothing to offer the fields of biology and medicine. When Pasteur announced that he was also having success in immunizing against anthrax, he was publicly challenged to demonstrate his vaccine in a farm environment with 50 sheep.

More luck in a second success

The public test was a resounding success. 25 unvaccinated sheep died, and 25 vaccinated sheep survived. It wasn't until many decades after his death that it was discovered that Pasteur used his rival's method for preparing the vaccine! His aides had prepared some vaccine from dead bacteria, and Pasteur decided to use that, rather than his own method, which only worked some of the time. So he ended up using a vaccine that worked in a way he couldn't understand.

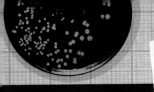

ANGELINA HESSE

DEVELOPS A BACTERIAL
GROWTH MEDIUM

1881

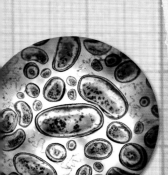

JELLING THE FUTURE OF BIOLOGICAL RESEARCH

Agar provides the perfect environment for bacteria growth.

MANY OF THE MOST SIGNIFICANT *advances in medical treatment have occurred because of the ability to conveniently examine and identify micro-organisms. In their native environment—say, inside your stomach—it's impossible to separate and investigate different bacteria. But for more than 100 years scientists have been able to grow bacteria under controlled conditions because of a chance observation that had its source not in the laboratory, but in the kitchen.*

Robert Koch, German bacteriologist.

THE GERM THEORY

In 1881, the germ theory of disease was on the verge of acceptance. The germ theory holds that many human diseases are the result of our bodies being invaded by invisibly small micro-organisms, bacteria, which make themselves at home inside our bodies. These bacteria not only move in, but they become so comfortable they multiply like rabbits—except much faster. As they acclimatize and make themselves comfortable, they make us rather uncomfortable. Sometimes these bacteria go far beyond causing discomfort, and they kill their human hosts.

Now we know this to be true, but in 1881 the story was just coming together. Scientists had found bacteria in sick or dead animals and people, but how to tell if they were responsible for the disease? Even harder, how to determine which one was responsible for a given disease? The trick was to take some of these recovered bacteria and incubate them—allow them to grow and reproduce—outside the body where they could be isolated and studied. Anyone who has forgotten a pot of chicken broth on the stove for a couple of days knows it's not too hard to grow bacteria. The problem is that the bacteria swim around in the broth and it becomes impossible to separate the different types in the soup.

BACTERIA SOUP

Gelatin! Gelatin can be boiled with all sorts of broths and it thickens into a nice semi-solid mass; so scientists of the time tried that. The problem is that bacteria which do well in the human body are active at our temperature, about 99°F (37°C). Gelatin does not do well at that temperature; it turns back into a broth. In fact, a warm day could (and often did) turn plates of bacterial cultures into soup.

TOP: *It is impossible to separate the different types of bacteria once they have grown in a soup.*

ABOVE: *Original culture plate on which Sir Alexander Fleming first observed the growth of* Penicillin notatum *in 1929.*

A HELPING HAND

That was the situation in 1881 when Walther Hesse started working with Robert Koch, one of the most important figures in the study of bacteria and disease. Koch asked Hesse to study airborne bacteria, but the gelatin just wasn't doing the job. Like many husbands encountering problems, Walther talked it over with his wife, Angelina. Unlike many such interactions, Angelina could bring her expertise to bear on the problem. She was not only Walther's wife, she was his laboratory assistant as well. She even allowed her kitchen to double as their laboratory. Which made it very convenient for her to respond to Walther's problem, because the answer was right at hand, on the pantry shelves.

Fannie Angelina Eilshemius was born in New Jersey in 1850, the daughter of Dutch immigrants. Her next door neighbors were also immigrants, by way of the Dutch East Indies, specifically, the island of Java. While in Java they learned of a remarkable gelling agent, distilled from seaweed. The gelling agent, called by its Malay name, "agar agar," is remarkable because when it returns from a boil down to room temperature (or well above, depending on the specifics of the preparation), it creates a semi-solid gel, and when the temperature rises, it stays as a gel. In fact it won't liquefy again until it reaches about 194°F (90°C)—well above the temperature needed to approximate conditions within a host's body. When Fannie married and settled in Germany she brought this knowledge with her, along with a supply of agar agar courtesy of her mother and the neighbors back in New Jersey. She used it for her fruit and vegetable preserves. Now she proposed another use: as a medium for bacterial growth.

AGAR

The chain of coincidence, from the East Indies to the laboratory of a pioneering microbiologist, had positioned Frau Hesse to make one of the most significant innovations in microbiological research. Agar (as it is usually called today) is a nearly ideal medium for bacterial growth. The consistency can be varied with concentration, the gel is stable, and it's translucent. Together these qualities mean that a plate of agar gel can be inoculated with a very dilute preparation of bacteria and separate colonies will spring up. The colonies, each started from just one or two bacteria, remain physically distinct; so it's easy to identify, study, and manipulate a single specific strain. For this application, and many more, Frau Hesse's agar gel is used today in virtually every biological laboratory.

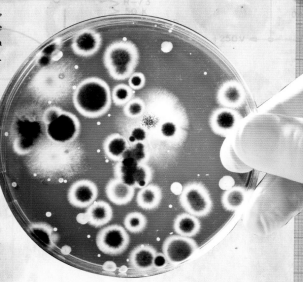

Agar gel is stable and translucent. When inoculated with bacteria, colonies are easily visible.

FACING PAGE: *Robert Koch, German physician and pioneer bacteriologist in search of the Rinderpest microbe at Kimberley. The scientist is shown at work, surrounded by another of his assistant's inventions: the petri dish (see page 93).*

Taking a peek

Robert Koch's laboratory was an incubator of discovery. Koch was the first to definitively link bacteria to a disease, proving that it wasn't "ill humors" or "bad air" that caused infectious disease. As we've just learned, Koch's drive for discovery was indirectly responsible for the development of the agar gel bacterial growth medium. Another assistant was responsible for an equally important development for microbiology.

Richard Julius Petri was another of Koch's assistants. One of his accomplishments was the design of a simple but efficient piece of laboratory glassware: a plate for holding the growth medium and protecting the gel from contamination. The plate was about 4 in (10 cm) in diameter, with a vertical lip rising about $2/5$ in (1 cm). The bottom plate would be filled with gel, then inoculated with the bacteria of interest, then covered with another glass plate made to fit over the lip of the bottom plate. The bacteria could grow in isolation, but still be viewed, and even taken under a microscope... which is why the petri dish is still seen in laboratories all over the world.

PHYSICS

MEDICINE

SEEING SKELETONS

NOVEMBER, 1895. *Winter darkness descended on the central German town of Würzburg, but it was even darker in the laboratory of Wilhelm Conrad Roentgen (Röntgen). He flipped a switch on a vacuum tube to check to see if he had effectively blocked all the light from leaking out of the tube. He had: the room stayed dark——except what was that? Feet away he saw a glowing spot that disappeared when he turned the tube off, then reappeared when he turned the tube back on.*

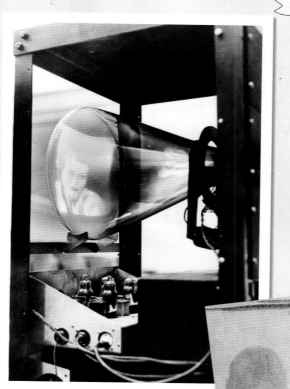

A picture of Joan Crawford as it appeared on a cathode tube after being televised from an adjoining room in the Franklin Institute.

Roentgen in his laboratory, 1901.

CATHODE RAYS

Roentgen, a Professor of Physics at the University of Würzburg, had previously turned his experimental talents on subjects ranging from the thermal conductivity of materials to the compressibility of gases and liquids. He was now turning those same talents to one of the hot topics in the world of physics at the turn of the twentieth century: what the heck were cathode rays?

Cathode rays were a new kind of matter—like a beam of light that could make the air glow, but unlike a beam of light, the cathode ray was quickly absorbed in air. Today we know that cathode rays are electrons generated by high voltage, but in 1895 no one knew what they were. Roentgen wanted to know; so he began by using a "Lenard Tube," a modified quartz vacuum tube that let cathode rays into the air. But after hours of work creating the vacuum, when the voltage

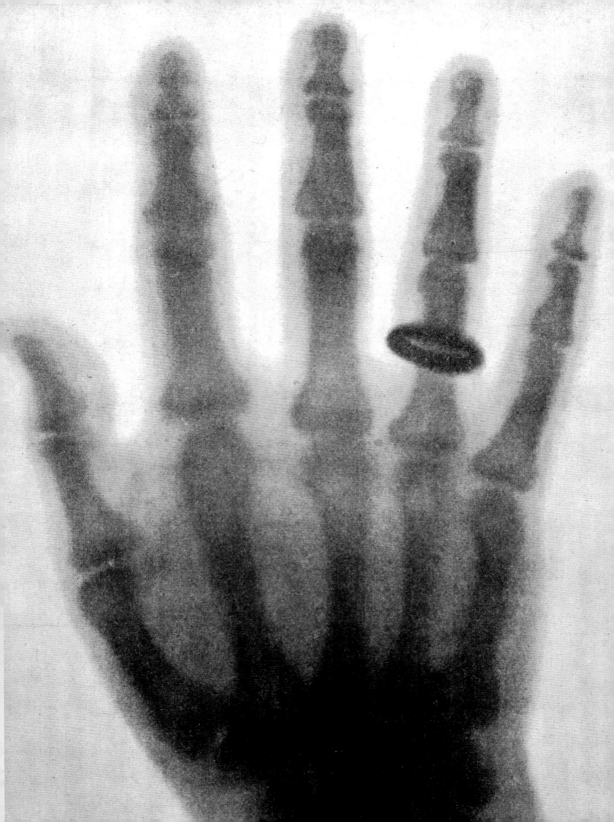

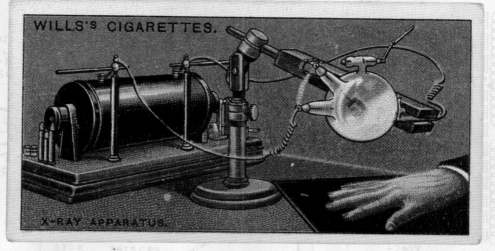

X-ray apparatus in 1895, the year X-rays were discovered.

was applied the seal would break after just a moment, and the tube would stop working. After suffering through this cycle of vacuum failure several times, Roentgen had had enough. He constructed his own special tube with one end of the quartz cylinder made as thin as possible. Then he cranked up the voltage, hoping he'd be able to force some cathode rays through the thin quartz into the air to be studied, just as in a Lenard Tube.

THE UNKNOWN QUANTITY

Roentgen prepared for the experiments. He knew cathode rays would just create a dim glow; so he covered his quartz tube in black paper. He also prepared sheets of cardboard coated with barium platinocyanide, a material that glows (or fluoresces) when cathode rays strike it. Both these effects, the fluorescence from the air and from the barium platinocyanide, are very dim; so he darkened the laboratory to view the effects. It was Friday night, his assistants had gone home, yet he was anxious to begin. But before he started his experiments, he had to be sure he had made the room dark enough to see the dim glow from the cathode rays.

He turned on the high voltage to his new tube and—what was that? 6 ft (2 m) away a cardboard sheet coated with barium platinocyanide was resting on the bench. When he turned on the tube, the sheet glowed back at him. He immediately repeated the test, placing the fluorescent sheet at different distances from the tube. A glowing spot appeared when the sheet was close to the tube, but also as it moved farther away. Everything then known about cathode rays—which we today know to be made of electrons—indicated that they couldn't travel very far in air. It didn't take Roentgen long to realize he was seeing something completely new, something unknown, which he labeled with the traditional name for an unknown quantity: "X."

Roentgen spent the next six weeks discovering as much as he could about the nature of the new radiation he had discovered. It's been written that it took 100 years for anyone to learn anything about the behavior of X-rays that Roentgen hadn't learned in those first six weeks.

FACING PAGE: *A photograph by Dr. Voller of Hamburg, taken using the Roentgen system, 1896.*

He learned that X-rays would easily go through cardboard, paper, rubber, wood… while holding up a metal weight in front of the fluorescent screen, Roentgen noticed the shadow of the bones within his own hand. He learned that X-rays could expose photographic film. He recruited his wife to hold her hand in front of a photographic plate and produced the world's first anatomical X-ray.

Years later, others claimed to be the first to discover X-rays, and they may be right, because others had seen the same kind of fluorescent flash that Roentgen had seen. They had the preparation and the opportunity—any one of several scientists could have recognized that the new radiation was much different than the cathode rays they were trying to study—but only Roentgen had curiosity and openness of mind to pursue the impressive set of measurements that were to give the world one of its most useful diagnostic tools.

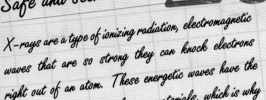

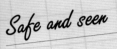

Safe and seen

X-rays are a type of ionizing radiation, electromagnetic waves that are so strong they can knock electrons right out of an atom. These energetic waves have the ability to zip right through many materials, which is why they're so useful, but also why they're so dangerous. Long-term exposure to X-rays distorts molecules in living cells, leading to cancer. Unfortunately the dangers were not known until many of the early X-ray practitioners grew tumors.

When Roentgen learned that photographic film could detect X-rays he built himself a darkroom so that he could do his experiments during the day. The X-rays were generated outside the room, came in through a small opening, and struck the film. Coincidentally, and luckily for Roentgen's health, he had built the walls of his darkroom out of lead, one of the most effective materials for stopping X-rays.

TOP: *A modern X-ray of a foot.*
ABOVE: *An X-ray tube.*
FACING PAGE: *Modern X-ray of a hand.*

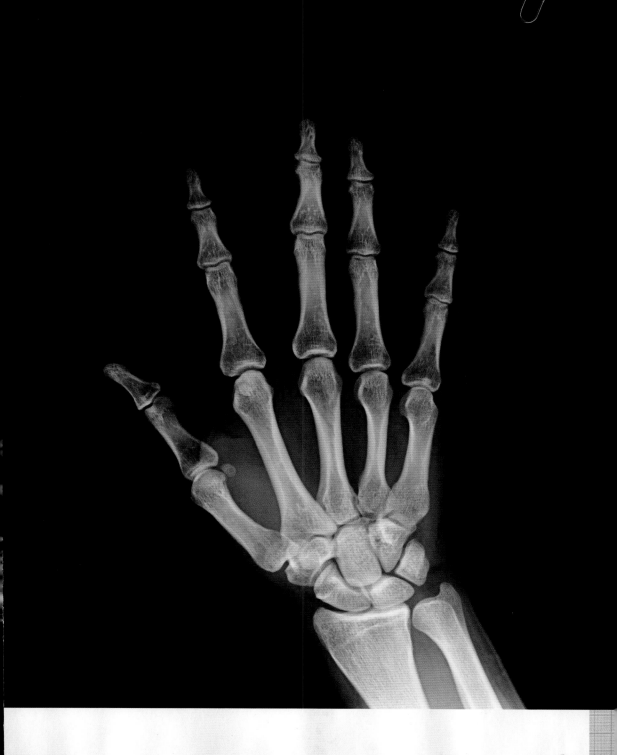

CHAPTER 3

LIVING BY THE RULES: THINKING OUTSIDE THE BOX

TOO TOUGH TO BREAK
Edouard Benedictus invents safety glass, 1903

MAKING IT UP
Leo Baekeland invents Bakelite, the first synthetic material, 1906

TURNING THINGS INSIDE-OUT
Ross Harrison develops tissue culturing, 1906

A MOLDY CHAIN OF CIRCUMSTANCE
Alexander Fleming discovers penicillin, 1928

In the years from 1900 to 1940 the trend toward scientific specialization was continuing. The educational framework and the structure of research was changing to keep pace. This period also marks the start of the transition from the model of the single researcher to the research team.

The idea of a truly lone scientist had nearly always been as much myth as reality. Even Newton, the perfect example of a solitary genius, acknowledged: "If I have seen further it is by standing on the shoulders of giants." In the century previous to 1900, though, there was still a fair measure of disjointed, separate research. Charles Goodyear and Ignaz Semmelweis, for example, both labored in isolation from their colleagues.

That was becoming much rarer. Research was becoming more interconnected, and research projects generally involved collaboration of one sort or another. The discovery of penicillin, for example, is one of the most famous accidental discoveries in history. Yet the story is not complete without mentioning three scientists at two separate universities who did their work more than ten years apart.

The educational system was evolving during these years. Many of the discoveries described in this chapter were made by researchers who had studied at more than one university or worked with teams of researchers at different facilities.

Corporate research and development facilities also became more prominent. To advance their technological development, companies needed to push the boundaries of science. Basic, fundamental science and applied research both had their place in the corporate environment, and both created environments where chance observation could lead to a big payoff.

This timeframe also saw the introduction of a new type of chance discovery. In the previous hundred years there were plenty of accidental discoveries where a researcher just "stumbled across" an interesting observation. Many times the scientists involved were doing research in a general field when they observed something that just didn't fit with what they were expecting. Malus, Oersted, Wöhler, Perkin, Roentgen; they were all trying to expand general knowledge about a topic. But after 1900 research became more targeted. Several of the stories from this time relate circumstances where researchers were on a quest for something specific when they found something quite different from what they were searching for. In a way, these discoveries are all the more remarkable for that. After all, these people had a job to do, but they ignored it, or perhaps just stretched the definition of their task to pursue a chance observation to its logical conclusion.

Even with the changes in training, research environments, and collaborative development, the usual elements of accidental discovery still needed to be present. Without preparation, a chance observation would mean nothing. And without the desire to pursue an unusual observation there would be no accidental genius.

BOUNCING AND RACING AROUND THE LAB
Wallace Carothers and Arnold Collins, Synthetic Rubber, 1930
Wallace Carothers and Julian Hill, Nylon, 1930

ASTRAL VOICES ON THE TELEPHONE
Karl Jansky opens the study of radio astronomy, 1931

COLORING THE FIGHT AGAINST DISEASE
Gerhard Domagk discovers the first antibacterial chemical, 1932

A BIG SLIP-UP
Roy Plunkett invents DuPont™ Teflon®, 1938

EDOUARD BENEDICTUS

INVENTS SAFETY GLASS

1903

TOO TOUGH
TO BREAK

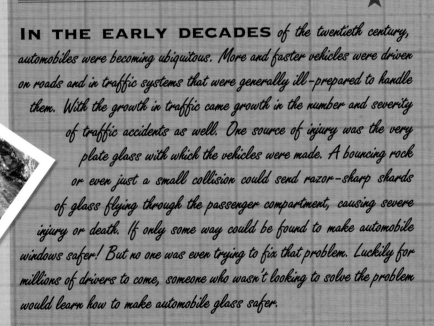

IN THE EARLY DECADES of the twentieth century, automobiles were becoming ubiquitous. More and faster vehicles were driven on roads and in traffic systems that were generally ill-prepared to handle them. With the growth in traffic came growth in the number and severity of traffic accidents as well. One source of injury was the very plate glass with which the vehicles were made. A bouncing rock or even just a small collision could send razor-sharp shards of glass flying through the passenger compartment, causing severe injury or death. If only some way could be found to make automobile windows safer! But no one was even trying to fix that problem. Luckily for millions of drivers to come, someone who wasn't looking to solve the problem would learn how to make automobile glass safer.

Nitrocellulose is made by treating cotton with acid.

INTRIGUED BY AN ODDITY

Edouard Benedictus was a French chemist and artist, born in 1878. In 1903 he was working with collodion, a thick, syrupy compound. Collodion was made by dissolving nitrocellulose in solvents. Nitrocellulose, or cellulose nitrate, can be made by treating cotton with acid. Collodion had already been used as a kind of binding glue to make celluloid, a durable, waterproof material used for men's shirt collars and cuffs. We don't know today what Benedictus was planning on doing with the collodion. We do know that through good fortune, he made two mistakes. First he forgot to put the stopper in a flask of nitrocellulose. Second, he knocked the flask off his bench where it fell to the floor and shattered into a thousa—No! It didn't shatter! Instead, the flask, although it broke, didn't separate into lots of little pieces; the pieces stayed stuck together. Benedictus was intrigued by the oddity. He labeled the flask with a note about

Two mistakes led to the discovery of safety glass: first, Edouard Benedictus forgot to stopper his flask, then he knocked it over.

It has been said that Eduouard Benedictus was driven to pursue the design of safety glass after he witnessed a gruesome traffic accident in which a young woman was badly cut by broken glass.

his observations. But his desire ended there, at least for the moment.

It wasn't until some time later that his thoughts returned to the case of the strangely shatter-resistant flask. Although the stories differ in some details, contemporary accounts say that Benedictus witnessed a particularly gruesome traffic accident in which a young woman was cut by broken glass. The accident he witnessed caused Benedictus to think of the flask that didn't shatter. He pulled the flask off the shelf and discovered that the solvents in the collodion had evaporated, leaving a nitrocellulose coating on the inside of the flask. The nitrocellulose had formed an adhesive layer holding together the shattered pieces. Reports

say that Benedictus came up with his "triplex" design within hours of the time he witnessed the accident. Triplex consisted of two sheets of glass pressed together with a layer of nitrocellulose in between. The laminated design was effective, but was it cost-effective?

Automobile manufacturers did not jump at the prospect of installing more expensive "safety glass" in their automobiles. The first customers were military. World War I had begun—a "chemist's war" where deadly gases were cannoned from one side to another. Gas masks were essential for the soldiers at the front, but the glass, as any glass, was breakable—with grave consequences attending a shattered lens.

Benedictus's safety glass had a customer.

THE POTENTIAL FOR DISCOVERY

Safety glass was not adopted by auto manufacturers for a few more years, and then only because of another chance occurrence. While Henry Ford was at work on the Model A design, one of his engineers was seriously cut by shattered glass as a result of an automobile collision. When it became personal, Ford understood how valuable safety glass could be (although a lawsuit may also have played a role). Ford first installed safety glass in 1919. Ten years later all Ford's models had safety glass windshields, and other manufacturers had joined the trend. In this accidental discovery, preparation and opportunity came quickly, but desire was slow to grow. Once the desire was present—first in Benedictus and then his customers —the discovery reached its potential.

The gases used in World War I created an urgent need for shatter-resistant gas masks.

Henry Ford was one of the first to recognize the value of safety glass.

The triplex safety glass design uses a three-layer system where layers are bonded using a central bonding agent made from an organic compound.

No substitute for safety

Benedictus's triplex safety glass design proved effective in reducing the incidence of injuries from glass shards, but it had another problem. Sunlight caused the nitrocellulose layer to turn yellow, reducing visibility. Cellulose acetate turned out to be a better choice. That in turn was replaced, but laminated safety glass is still made with a triplex design. Now the central bonding layer is typically provided by another organic compound, polyvinyl butyral (PVB). In addition to providing shatter resistance, PVB-laminated windows also provide good sound insulation.

In addition to the laminated triplex design, safety glass is also available in tempered glass. The tempered glass version is created by heating one side of a glass sheet, then cooling it rapidly. The thermal treatment creates stress within the glass. The lines of stress are oriented perpendicular to the two surfaces; so when tempered safety glass breaks, it shatters into many tiny squared-off pieces without the deadly shards that are created when untempered glass is broken.

Both types of safety glass are widely used today. The laminated design is generally preferred for windshields, because the glass can be partially damaged and still retain its visibility beyond the damaged area, while tempered glass disintegrates into thousands of tiny chips. So Benedictus's accident is still paying dividends in health and life. Good thing he was a little clumsy!

LEFT: Tempered safety glass.
RIGHT: An organic molecule (the base unit of the central bonding agent now used in safety glass production).

BELOW AND BOTTOM:
Because it is a good insulator, Bakelite has been used to make all manner of materials from volt meters to telephone wires.

LEO BAEKELAND

INVENTS BAKELITE, THE FIRST SYNTHETIC MATERIAL

1906

MAKING IT UP

IN THE EARLY TWENTIETH CENTURY

the burgeoning electrical industry was driving demand for raw materials, both metals and insulators. Metals were necessary for cables. Porcelain or glass insulators mounted on poles supported those cables. Insulation wrapped the cables. Early cable insulation was made from shellac, a product created from a resin secreted by the Lac beetle of South Asia. Refining shellac is a labor-intensive process, as well as being rather limited by the amount of resin a little beetle can make. Another source of cable insulation was needed, and chemist Leo Baekeland was determined to find it. For three years he tried, unsuccessfully. Finally, after three years of failure, he failed again. But this failure became an unexpected success that changed just about every manufacturing activity in the world.

The original Bakelizer, used by Leo Baekeland and his co-workers from 1907 to 1910 to form Bakelite by reacting phenol and formaldehyde under pressure at a high temperature.

Leo Baekeland was born in Ghent, Belgium, trained as a chemist at Ghent University, then in 1889 was offered a traveling scholarship to the United States. He never moved back. He was determined to invent something that would provide a reasonable income for his wife and children. In the early 1890s he found success with a new type of photographic paper that could be printed with artificial light, rather than with exposure to sunlight like the photo paper then in use. He formed a company to make his new "Velox" paper. Then Eastman Kodak bought Baekeland's company, providing him not only enough money to set his family up in style, but also enough to make himself a private laboratory.

Bakelite was initially used in radio and telephone casings because of its electrically non-conductive and heat-resistant properties.

A LAB OF LOVE

Baekeland built his home lab out of love, not necessity, because he had received a faculty appointment at Columbia University, where he worked for the next 17 years. The University appointment still allowed him the freedom to pursue his own researches; so he divided his time between his lab at Columbia and his lab at home. He was satisfied with both the wealth and the fame he had achieved. In fact, he thought perhaps he was a little too wealthy—he was making more than the "reasonable income" he had set out to achieve, and he couldn't quite reconcile himself to accepting the style of living he was now expected to maintain. He preferred a frugal lifestyle; so he didn't want more money, but he still wanted to discover and create.

He was about to create something that changed history.

Baekeland turned his attention to the problem of wire insulation. Earlier chemists had discovered that mixing an organic molecule of a type called a phenol with one of a type called an aldehyde could create a resin-like compound. He tried heating up various phenols and aldehydes together, but when they cooled they were either too hard or too soft. Then he realized he could change something else. Perhaps it wasn't a problem with which compounds he was mixing, but the way he was mixing them. He built what he called a "bakelizer," an iron container that allowed him to regulate the temperature and the pressure of the reactions (see illustration, p. 109). Then he mixed carbolic acid, a coal tar derivative, with formaldehyde, a wood alcohol product, and put the two chemicals in his new device.

Something heat resistant and non-conductive was needed to insulate wires—Leo Baekeland turned his attention to this.

Bakelite was the first plastic to be used for making radios, and was ideal for the Art Deco-style designs of the 1920s and 1930s.

A Bakelite radio from the Bakelite Museum in Somerset, England.

SOMETHING COMPLETELY NEW

Baekeland ended up with a transparent, insoluble resin that was no good as a replacement for shellac. It wasn't a liquid, but a hard, compact mass that he described as something like hard rubber or ivory. Then it occurred to him that the homogeneous resin could be molded into useful shapes. He was right. He called his new product "Bakelite."

He had invented the first plastic. Almost overnight it was used to make fountain pens, billiard balls, toothbrushes, radios, telephones, ashtrays, airplane propellers—hundreds of different products. Successive years would see new plastics developed. All trace their lineage back to Bakelite.

But even more important than that, Baekeland had created the first fully synthetic material.

He took two unrelated compounds and created something completely new. With that creation, a world of possibility opened. Manufacturing was no longer limited to what could be found and modified, now materials could be made to meet the requirements of an advancing society.

In a sense, this epitomizes the law of unintended consequences. Baekeland had prepared himself with years of training so that he could satisfy his need to discover and create. He didn't envision himself creating a new world, but he created it nonetheless. Imagine a world without plastics. Thanks to Baekeland, you don't need to.

Creating a new reality

In 1906 Leo Baekeland presented his new Bakelite——the full chemical name is polyoxybenzylmethylenglycolanhydride——to the Chemist's Club in New York City. The assembled scientists were treated to a demonstration of the light, durable, moldable material created entirely in the laboratory. They rose to their feet and applauded. Baekeland had brought a new material into the world, sure, but to these chemists he brought the world into an era of new materials. If something didn't exist, they could create it.

A sense of the optimism this engendered can be found in a New York Times article of August, 1926. The article reported that scientists promised "that substitutes will be found for everything that man needs if natural supply fails." The article discussed a scientific conference where chemists envisioned a world "freed from the tyranny of raw materials," with synthetic replacements for wood, fabric, flooring, pharmaceuticals, even food! The amazing thing is that the chemists

FACING PAGE:
Leo Baekeland, US chemist (1863–1944) who invented Bakelite and patented it in 1906, pictured in his laboratory.

were right. Before there was even a well-defined path to most of these materials, they knew what was possible. Baekeland's invention had burst through a barrier, and there was no turning back.

Ironically, the plastics that Baekeland metaphorically fathered have become the symbol for the profligate, wasteful lifestyle he disdained. Baekeland lived more simply than his means would have allowed, preferring frugality over conspicuous consumption.

ROSS HARRISON

DEVELOPS TISSUE
CULTURING

1906

TURNING THINGS INSIDE-OUT

IN THE EARLY 1900s, *Ross Harrison was a zoologist studying fish and frog embryo development. The work he had chosen dumped him right in the middle of one of the biggest biological controversies of the day: how does the nervous system develop? Along the way to answering the question he created a method — without intending to — that immeasurably helped the understanding and prevention of disease.*

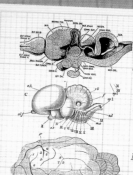

Shark Brain

Pigeon Brain

Dog Brain

Basic nervous system of different animal species.

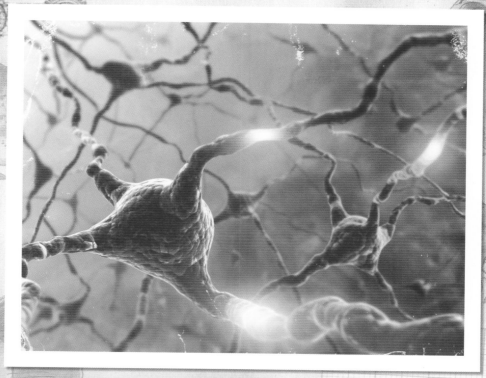

Inside the brain: two neurons transmit information as part of the nervous system.

A UNIQUE PERSPECTIVE

Harrison received his Ph.D. in 1894 from Johns Hopkins University and an M.D. from the University of Bonn in 1899. At the same time he did field and laboratory work in zoology, studying a variety of marine and terrestrial creatures. Harrison's combination of medical and zoological training gave him a unique perspective. In 1906 he turned his attention to the nervous system, the brain, spinal cord, and complex network of nerve fibers that stretch throughout the body.

As a vertebrate embryo is developing, neurons first appear in a clump, called a neural tube, that develops into the brain and spinal cord. But soon afterward nerve fibers appear throughout the body. But what would happen if the signals got

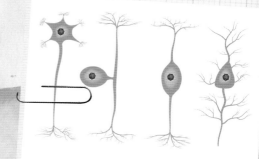

Annotated diagram of neurons and their various parts at work in the body.

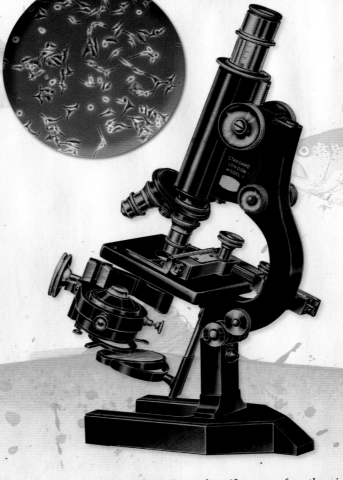

RIGHT: *HeLa (cervical cancer) cells in culture—one of the most commonly used in biological research.*
BELOW: *Close-up photo of frogs' eggs.*

mixed up, if the nerves got wired the wrong way during development? What if, say, a fish scraping its right side against a coral outcrop felt the scrape on its left side? Or if it tried to swish its tail fin and instead swung its dorsal fin? Such a fish would not live too long. Somehow the nervous system needs to be organized.

COMPETING THEORIES

There were three competitive theories about how the nerve fibers end up connected to the correct part of the brain. First, that the cells along the nerve paths create the nerve fibers as one interconnected mass. Second, that the embryo contains a network of fine threads at its very earliest stages, and the nerve fibers are converted from the pre-existing fine threads. Third, that the nerve fibers grow as

extensions of specific neurons from the spinal cord. The first two proposed mechanisms are attractive because the plan for guiding nerve growth is built right into the developing organism, but they imply that the nerves aren't composed of individual cells, as is every other organ in the body.

If nerve fibers grow as a single mass along pre-established paths then the brain can be pre-wired. That is, before any of the nerve fibers are connected, the brain would contain exactly the information it needs to, say, blink the left eyelid. If the nervous system "grows itself" then the brain would have to learn exactly which nerves are connected where—the brain would need built-in mechanisms to teach itself. What we learn about animal brain development will guide our understanding of human development and learning. That's why the question of nerve development was so important.

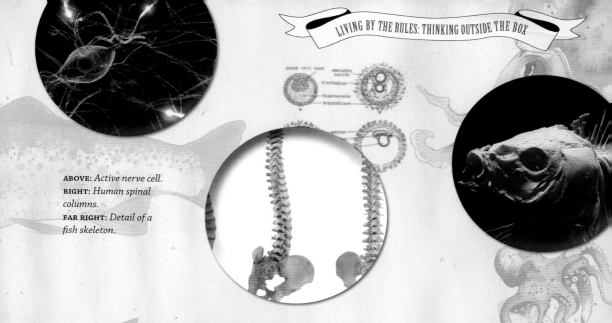

ABOVE: *Active nerve cell.*
RIGHT: *Human spinal columns.*
FAR RIGHT: *Detail of a fish skeleton.*

Zoologists of the day conducted their science as a kind of passive investigation. To answer the question about neural development, scientists examined the nerve fibers in animal embryos at various stages of development. The problem was that scientists could look at exactly the same nerves and have varied explanations. Something different needed to be done.

Harrison had experience transplanting tissue in frogs and he was trained as both a medical doctor and a researcher. Harrison was perfectly prepared to "do something different." He removed the neural tube from frog embryos and no nerve fibers grew. He removed the spinal cord and replaced it with connective tissue, and nerve fibers grew from the brain through the new tissue. He even replaced the spinal cord with a blood clot, and nerve fibers grew through the clot. The evidence seemed clear. Nerve fibers didn't grow throughout the body, they stretched out from the neural tube. Yet there were still arguments. So he took his experiment one step farther.

He removed a nerve bundle from a frog embryo and, rather than transplanting it into another embryo, he put it on a glass slide in a drop of frog lymph. He watched the prepared slide for as long as a few weeks. The nerve fibers grew. There were no other body cells for the fibers to grow from,

only the neurons. Now there could be no question: Nerve fibers grew from neurons.

Harrison published his work in triumph. He had succeeded in doing what he had set out to do. So where is the accident?

A SIGNIFICANT QUESTION

Harrison had done much more than answer an admittedly significant question. He had taken cells out of a body and kept them alive. Suddenly processes that had been hidden inside an organism could be studied under a microscope. He was not intending to, but Harrison had created an entirely different way of studying life.

Harrison was done with this technique, but once he had shown it was possible, others developed it into a valuable method. Prior to tissue culturing, viruses needed to be maintained in living animals. Now tissue cultures are an absolutely essential tool used to investigate the cellular and molecular mechanisms of cancer, AIDS, and other diseases. A recent book selected the development of tissue culture as one of medicine's ten greatest discoveries because it has advanced the understanding of the mechanisms of disease more in the past 50 years than was learned in the previous five thousand.

Immortality in a dish

The tissue culture method that Ross Harrison pioneered allowed him to dramatically demonstrate the mechanism of nerve growth. Harrison's tissue-culturing method, called "in vitro" after the Latin words meaning "in glass," was limited to growing frog tissue. It was more difficult to grow tissue from warm-blooded creatures, but more useful as well. Alexis Carrel and his assistant Montrose Burrows took the next steps. Carrel famously kept some chicken heart tissue alive by periodically cutting it in half and replenishing the chicken plasma and embryonic fluid used to nourish the tissue. The constantly dividing cells in the tissue were reportedly kept alive for 34 years, outliving Carrel himself.

Unfortunately for the growth of scientific understanding, the tissue was not the living descendant of the initial sample, but a result of contamination from new cells that had been introduced with each refill of chicken embryonic fluid. The contamination was not revealed until after Carrel had died. It's not known whether Carrel knew his "immortal" chicken cells were nothing of the sort. Carrel's results had created confusion because other

Alexis Carrel was a major in the French Army Medical Corps during World War I.

researchers had found that ordinary healthy cells can only be kept alive up to about 50 cycles of division.

On the other hand, cancerous cells are hardy and aggressive and some can stay alive indefinitely. A particularly famous cell line was taken from cervical cancer cells of a woman by the name of Henrietta Lacks in 1951, the year she died. Billions of her cells are still living in laboratories around the world. In fact, these cells are so hardy that if just a few cells accidentally spread to another cell culture, the HeLa cells (see image on page 116) will take over. Just recently it was discovered that what were thought to be independent cell lines had been contaminated and taken over by HeLa cells. For more than 50 years HeLa cells have played a significant role in the study of disease, a role only possible because of Harrison's accidental development.

Alexis Carrel led a conference in a Parisian hospital in 1913 on the transplantation of organic tissue.

ALEXANDER FLEMING

DISCOVERS PENICILLIN

1928

A MOLDY CHAIN
OF CIRCUMSTANCE

WITHOUT EXAGGERATION, penicillin has been called the most powerful disease-fighting chemical in existence. Within short months of its initial availability in the midst of World War II, thousands——tens of thousands——of lives had been saved. Those lives and the millions that followed in the next decades all owe themselves to a remarkable string of chance occurrences. At the heart of the discovery is a key observation at a critical time.

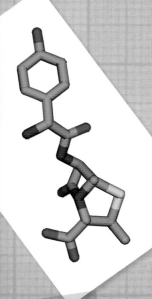

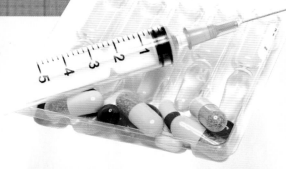

LEFT: *The molecular structure of Penicillin.*
RIGHT: *"Mold juice filtrate," as it was called by Alexander Fleming, is extracted using a syringe. It was later identified as a chemical extract of* Penicillium notatum.

Fleming at work in his laboratory.

A POSITION ELSEWHERE

In 1928 Merlin Pryce was a research assistant at St. Mary's Hospital in London in Alexander Fleming's laboratory, working to evaluate the stability of a particular strain of infectious *Staphylococcus* bacteria. Pryce's work was not complete when he was offered a position elsewhere. To study such bacteria, biologists fill a small circular glass plate—a Petri dish—with a nutritious gel that feeds the tiny microbes. Fleming had prepared a set of dishes, introduced bacteria on the gel, then covered the dishes, a process called "culturing."

Although some details are hazy, it appears as if Fleming then left on vacation, neglecting to put the cultured plates in the incubator, which keeps the temperature just at the right point for bacteria to grow. So instead of being in the incubator at a temperature around 99°F (37°C), the plates were kept at room temperature for a few weeks. Perhaps when Fleming came back he recognized the mistake he had made, or perhaps he just gave the cultures a cursory inspection, but it appears certain he tossed the dishes into a sterilizing bath. Then Pryce stopped by for a visit, and Fleming pulled one of the dishes from the top of the stack—a dish that had been atop other dishes and stayed clear of the sterilizing solution. Fleming wanted to show Pryce what he'd been doing with *Staphylococcus* since Pryce had gone, but when he picked up the plate he said, "that's funny."

SOMETHING IN THE MOLD

The plate had been contaminated by mold, perhaps carried through the air from a laboratory on the floor below where a colleague was studying allergies. Mold contamination wasn't too unusual, but Fleming noticed that the bacterial colonies near this mold had died while the colonies further away were thriving. Something in this mold, later

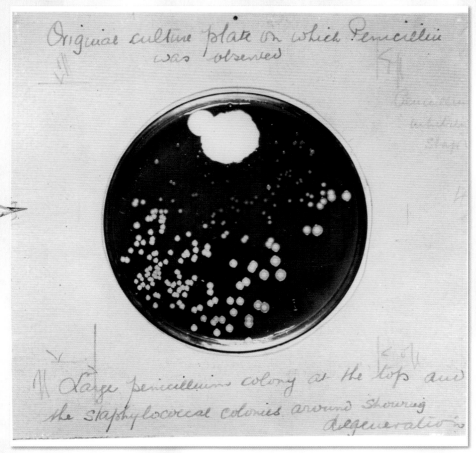

The original culture plate on which Penicillin notatum *was first observed in 1928.*

identified as *Penicillium notatum*, was killing staphylococcal colonies. Fleming extracted the compound in what he initially called "mold juice filtrate," and eventually named penicillin. The new compound effectively killed infectious bacteria while not affecting the health of white blood cells. He injected the new compound in mice and rabbits and they showed no ill effects. He even showed that the new compound could effectively treat eye infections. What he didn't do is see if penicillin could eliminate illness when injected into a laboratory animal. That test remained for others to do. 12 years later, others did, and Fleming's discovery reached fruition.

JUST THE RIGHT MOLD

Elements of the story are in dispute (there are competing recollections: Did Fleming leave the plates during a summer vacation? Did the mold come from the lab below? Was the critical plate really rescued from the disinfectant bath?) But it's indisputable that chance played a huge role in the discovery of penicillin. To begin with, the mold is very rare, and the chances of the right plate catching a spore from just the right mold were pretty small—none of the other molds from the downstairs laboratory showed any antibacterial activity. The growth conditions needed to be perfect: the antibacterial action would only work if the mold got a chance to grow at a low temperature before the bacteria grew at a higher temperature, and the weather just happened to provide the right conditions. And once the penicillin had been extracted, if Fleming had checked for safety in guinea pigs instead of mice or rabbits, he would have discovered that penicillin was toxic to guinea pigs, probably ending interest in the compound right there.

CHANCE OBSERVATION

Although uncertainties surround the story, there are some undeniable facts. Fleming was excited enough about his initial chance observation to show all his colleagues the mold-contaminated culture and take a photograph of the dish. He also made certain that he captured a sample of the mold and maintained a supply of healthy *Penicillium notatum*, a supply that ended up seeding development of the most effective medicine in history.

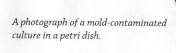

A photograph of a mold-contaminated culture in a petri dish.

ABOVE: *The experimental plant for extracting penicillin that was set up by Dr. A. G. Sanders, one of the penicillin research team at Oxford University.*

RIGHT: *Sir Ernst Boris Chain (1906–1979), British biochemist, born in Germany.*

Howard Florey, Australian pharmacologist and pathologist, injects penicillin into the tail of a mouse c.1940.

The promise is realized

Although Fleming's discovery was an essential element of the development of penicillin, it is inarguable that the drug would have remained undeveloped for much longer if Howard Florey and Ernst Chain of Oxford had not become interested. They converted many of Fleming's casual observations into concrete tests. Then they did a series of animal tests that demonstrated penicillin could be used as a systemic antibiotic: it could be injected into an animal and fight an existing infection. Finally, they developed methods for producing and purifying penicillin that proved the promise was there. The argument over the relative contribution of the scientists involved is still active, but it appears certain that Florey and Chain did experiments that Fleming could have done, if he had possessed the desire. In this case, Fleming recognized the significance of his observation, while Florey and Chain embodied the desire to understand all the implications.

WALLACE CAROTHERS
AND ARNOLD COLLINS

SYNTHETIC RUBBER

1867

BOUNCING AND RACING AROUND THE LAB

★

FOR A FEW YEARS *the DuPont Corporation's Pure Science Section in Wilmington, Delaware was doing fundamental scientific research as important as that done anywhere in the world. In a brief span in 1930 that lab's breakthroughs led to products that would be used by literally billions of people. It all started with Wallace Carothers and a controversy about very large molecules.*

VERY LARGE MOLECULES

Wallace Carothers began his academic career by studying accounting, taking classes at the commercial college where his father taught. Carothers then attended tiny Tarkio college in Missouri, where he taught business classes. He soon found himself drawn to courses in chemistry. He demonstrated such a deep understanding of the subject that he was asked to take over teaching duties when the chemistry professor left the college. Carothers's reputation grew, and he was invited to take a professorship at Harvard University. Soon after he started at Harvard he was asked to work at DuPont's new Pure Science Center. Carothers refused and was asked again. That cycle repeated several times until—after he was assured he would have the freedom to do fundamental science—he accepted.

Several monomers held together by chemical bonds form a polymer.

IDENTICAL BUILDING BLOCKS

Why was Carothers in such demand? He had an excellent grasp of the burgeoning field of organic chemistry. Organic chemistry was originally defined as the study of molecules that are only produced in living organisms, but became better defined as the study of complex molecules containing carbon atoms. On the face of it, there is no reason why organic chemistry should be any different from inorganic chemistry. For example, a whole separate type of chemistry could be defined for molecules that contain nitrogen or potassium or lead or any other atom. What's so special about organic molecules, complex molecules that contain carbon? The answer is not that organic molecules are interesting simply because of where they come from or where they naturally appear, they're interesting because of different ways they are put together. Molecules with identical numbers of carbon atoms can have different shapes, therefore different chemical properties. And molecules that are almost the same, say, one atom different out of 40, can seem completely unrelated.

Wallace Carothers understood organic chemistry as well as anyone, and that's why DuPont wanted him. There was so much possibility with organic compounds because small changes could make a big difference. Management at DuPont knew that there had to be economically valuable organic compounds yet to be discovered, and they knew Carothers was the man to get things started.

The first question Carothers needed to answer was about very large molecules— called macromolecules— which we now know of as polymers. A "typical" molecule, take table sugar, for example, has a molecular weight of about 343. Large polymer molecules, such as those that comprise rubber or silk, have molecular weights as high as 4,000. "Typical" molecules are held together with one or another form of chemical bond. In the 1920s many scientists supposed that "typical" molecules were held together with "typical" forces while the unusually large polymer molecules must be held together with unusual forces. German chemist Hermann Staudinger disagreed. He maintained that polymers consisted of long chains of smaller molecules held together with ordinary forces. Carothers was to prove him right.

Carothers knew that alcohols, molecules with an "extra" oxygen-hydrogen (OH)

Hermann Staudinger, German chemist, 1953.

The shape and texture of different substances comes down to their different chemical properties, caused by the type of bonds and different atoms in each building block.

group at the end, would easily bond with certain acids that ended with an "extra" carbon-oxygen-oxygen-hydrogen (COOH) group. When acids and alcohols come together, the "extra" parts connect through a carbon-oxygen-oxygen, leaving an extra two hydrogens and an oxygen: H_2O, or water. This chemistry was well-understood: an acid and an alcohol come together to make a molecule called an ester, with an extra water molecule on the side. But Carothers didn't want to make a molecule that was twice as big as a "typical" molecule, he wanted to make molecules much bigger. So he started with organic molecules that had an alcohol group hanging loose on one end and an acid group hanging loose on the other end. He figured that

each ester molecule could then attach to another ester molecule at each end and that he could make a long chain of many esters. He succeeded, creating a class of chemicals named using the Greek root for many: polyesters. Polyesters are just one example of a type of molecule built from a bunch of identical building blocks called monomers. When monomers are put together, they're called polymers. Carothers didn't invent polymers—rubber and silk are just two examples of naturally occurring polymers—but he was uniquely prepared to take advantage of a couple of laboratory accidents. His preparation and desire changed some chance observations into billions of dollars of manufactured products.

The benefits of impurity

Vulcanized rubber had been commercially available for half a century, and it was being used for everything from inked stamps to automobile tires. It was a lucrative market, and it all hinged on the availability of raw latex from rubber trees. The DuPont Company saw this as an opportunity for a synthetic product to capture a large market. Rubber is a polymer of isoprene, a molecule with four carbon atoms and five hydrogen atoms. In the mid-1920s chemists in DuPont's applied division were working on a method to make isoprene by connecting up acetylene molecules, which are composed of two carbon atoms and two hydrogen atoms. They figured they could join a bunch of acetylenes together mixed with some reactive hydrochloric acid, and the acetylenes would use the hydrogen atom from the acid to connect themselves into a long rubber-like molecule. But they couldn't do it. They ended up with just a bunch of short molecules. So when Carothers opened his polymer research group at DuPont's Pure Science Center, the applied chemists turned the problem over to him.

Carothers assigned Arnold Collins to the problem of making a poly-acetylene molecule. Collins was an excellent chemist, and he was able to separate perfectly pure liquids composed of molecules made from two or from three acetylene molecules, but he didn't make any molecules bigger than that. During his process, Collins had also separated out a yellowish liquid contaminant, a material that wasn't the joined-together acetylene he was looking for. Instead of throwing away the flask that contained the contaminant, Collins left the impurity sitting on his bench over the weekend.

Synthetic rubber is now part of our daily lives—the soles of shoes are just one example of its use.

When Collins came back on Monday, the yellowish liquid had turned into a solid mass. Collins knew that it had solidified because the individual molecules had joined up with each other all by themselves, that is, they had spontaneously polymerized. When he found this unexpected glob on his bench, Collins did what any self-respecting chemist would do. He stuck a wire into the flask, fished out the solid chunk, and played with it. He picked it up and squeezed it. It deformed and returned to its original shape. He tossed it on the floor. It bounced. At first glance, he had succeeded in making an artificial rubber! But had he? And, if he had, how?

Carothers led the effort to characterize and analyze the new compound. It was, like rubber, an elastomer. That is, it was a long chain polymer that could stretch to at least twice its starting dimensions and then return to the size at which it had started. It was better than rubber in that it was more resistant to degradation from petroleum products, other chemicals, heat, and ozone. So it was an excellent candidate for a synthetic rubber.

But what was it? It turns out that the new molecule in the yellowish liquid was almost exactly like isoprene, the monomer that combines to make rubber. The difference? One of the carbon atoms and a few hydrogen atoms were replaced by a chlorine molecule. This was completely unexpected. In fact, if a different acid had been used to help combine the acetylene molecules, the new synthetic rubber would not have been created. That was the first chance occurrence. But if Collins had just discarded the waste product, the thing he wasn't trying to make, he wouldn't ever have seen the spontaneous polymerization and wouldn't ever have known he had made a synthetic rubber.

And what a synthetic rubber! After the synthesis process was refined and developed, the new product found applications from O-rings to electrical insulation. It also has unique properties that make it just right for making wetsuits. In 1937 DuPont began marketing the material as neoprene. Today 300,000 tons of neoprene are made every year.

A GUIDED DEVELOPMENT

After its initial discovery, the story of neoprene moves from the arena of accidental discovery to the category of guided development. Within DuPont it also moved from the Pure Science Center to the applied divisions, and Carothers was no longer directly involved in the development of neoprene.

It's all right; he was busy with other things.

WALLACE CAROTHERS
AND JULIAN HILL

NYLON

1930

Alluring...Enduring...

Wolsey

nylons

A HECK OF A STRETCH

As noted previously, Carothers was deeply interested in the science of polymers. It could be said he was in an unofficial competition to make the longest molecules possible. He had moved past molecules with a weight of 4,000 and was routinely making polyesters with molecular weights around 5,000 to 6,000. But he seemed to have run into a roadblock. When two esters join together they release a water molecule, but the opposite reaction can also happen. When two joined esters are in the presence of a water molecule, there is a chance the esters will split and the hydrogen and oxygen from the water will rejoin the individual molecules. Carothers designed a "molecular still" to pull the water out of the solution. Then he charged Julian Hill with making the heaviest polymer possible.

Hill prepared the initial compound on the laboratory bench, then "finished off" the reaction in the molecular still, where the water could be removed. The resultant liquid was thick and viscous, a sign that it consisted of long polymers. In fact, Hill's polyester had a molecular weight of about 12,000, well heavier than any molecule that had been previously synthesized. When Hill was transferring the material from the molecular still, he noticed it was stringy—it could be stretched into fibers.

Hill, like Collins, decided to take the new material and play. Hill organized an impromptu contest. He passed out glass rods to a handful of

Synthetic polyester fiber was revolutionary for the world of textiles.

fellow chemists. They all dipped their glass rods into the syrupy liquid, then they ran down the hall trailing long, thin strings behind them. They were "cold-drawing" the fibers. The winner was whoever could make the longest string. The winner is not recorded, but, in a sense, everyone who has ever worn clothing made from synthetic fibers became a winner on that day. Hill noticed that the fibers pulled from that liquid mass became thinner and thinner—almost like silk—yet became stronger the more they were stretched.

Hill and Carothers examined the "cold drawing" process. They determined that pulling lined up all those long polymer molecules along the direction of the pull. In addition, the pulling process also established "crosslinks," chemical bonds between the different long strands. The result was the first synthetic polyester fiber, the ground-breaking result that revolutionized textile production.

Under the direction of Carothers, two playful experiences in the lab were converted into products that changed the world.

Links and more links

Polymer molecules are formed when individual molecules—the monomers—join together end-to-end to make one long molecule. In latex (the raw material in rubber), for example, isoprene molecules join together to make polyisoprene. The polyisoprene molecules are like long strings, but rather than being straight, the polyisoprene strings are slightly coiled. The long molecules "grab against" each other, making the material thick. And because the individual molecules are coiled, they can stretch and return to their original shape. So a ball of latex is like a hunk of sticky strings all interspersed with each other. And, just like that hunk of sticky string, a ball of latex can "melt" as the long molecules slide along each other, or the ball can crack when it's cold and the strings become a little less "sticky."

Goodyear fixed this by adding sulfur to rubber. The "vulcanization" process was really just a matter of connecting the strings to each other, not end-to-end, but side-to-side—crosslinking. The sulfur atoms hooked neighboring strings to each other, giving rubber enough stability to keep its shape.

Carothers' and Collins' polychloroprene—the synthetic molecule that became neoprene—was born already vulcanized. That is, the crosslinks already existed, making the product into a stable form. Unfortunately, as anyone who has tried to recycle tires can say, it is nearly impossible to change the shape of a piece of vulcanized rubber. Luckily for Carothers and DuPont, it was possible to stop the polychloroprene process in the middle, making a workable raw material. And luckily for scuba divers everywhere, one property of the material is its ability to form little pockets that can hold air.

Neoprene wetsuits are good insulators because they hold a layer of air between the diver's body and the bulk of the water.

THE HOT NEW MATERIAL

The polyester that Carothers and Hill discovered in 1930 did not end up being commercially valuable. The cold-drawn polyester fibers weren't long enough to be a practical replacement for silk, and, even worse, dry cleaning chemicals or even the heat of ironing would cause the fibers to disintegrate. Carothers knew that polyesters, joined with a carbon-oxygen-oxygen-hydrogen bond, melted at a lower temperature than polyamides, which are very similar except that they're held together with a carbon-oxygen-nitrogen-hydrogen bond. He tried to make some polyamides, but the problem was that their melting point was too high. The polymer needed to be melted so that it could be drawn into a fiber, but the polyamides had such a high melting point that they generally started breaking apart before they melted.

So for some time Carothers concentrated on trying to make more robust polyesters. It

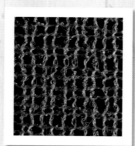

Nylon stockings are made from pulled polymer molecules. Cross-links formed between the molecules help it stay strong and stable.

wasn't until 1935 that he fully investigated a polyamide he had first made in 1930. He found, to his surprise, that it remained stable even up to its high melting point, and that it could be drawn into thin, strong, long fibers. The new material, which DuPont advertised as being made from "coal, air, and water," was nylon. It was a material of such remarkable properties that it changed the nature of the textile business.

With such incredible successes behind him, it would be understandable if Carothers had succumbed to an over-developed sense of self-importance. Instead, he wondered if the work he had done had any value at all. He became so depressed over his apparent "failure" that he took a fatal dose of poison while he was still at the height of his creativity.

KARL JANSKY

OPENS THE STUDY OF
RADIO ASTRONOMY

1931

ASTRAL VOICES ON THE TELEPHONE

IN 1927 THE AMERICAN TELEPHONE
and Telegraph Company (AT&T) initiated transatlantic telephone
service via a radio link between London and the United States. The
system could handle one call at a time, and the connection cost
a mere $75 for the first three minutes. At that price, it's no
wonder that users demanded at least a decent quality connection.
But calls were often interrupted by static, and AT&T wanted
to figure out how to stop the interruptions. They hired Karl
Jansky, a 22-year-old physics graduate from the University
of Wisconsin, to find the sources of static. Jansky found
the sources, but his investigation took him far from anything
AT&T could do anything about. It took him to the stars.

By correlating weather reports with bursts of static, Karl Jansky was able to attribute static electricity to distant thunderstorms.

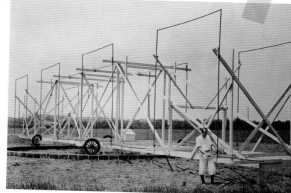

Jansky with his directional radio aerial system, the precursor of today's radio telescopes.

AN UNGAINLY CONTRAPTION

Karl Guthe Jansky was born in the Territory of Oklahoma in 1905. His father was the Dean of the College of Engineering at the University of Oklahoma, and encouraged Karl's interest in the rapidly growing field of radio technology. Karl got his Bachelor's Degree from the University of Wisconsin and joined AT&T's Bell Labs soon afterward. His first task was to build an antenna in what was then the remote area of Holmdel, New Jersey. The site was chosen to be far from city interference generated by electronic devices. Jansky built an ungainly 100-foot-long (30-meter-long) contraption of brass and steel rotating on four wheels pulled from an old Model T. Rotating the antenna allowed him to detect signals from different directions. Then he measured.

Jansky found that local thunderstorms generated loud bursts of static. He also heard less emphatic bursts of static. By correlating with

Jansky stands in front of an antenna at the Bell Telephone Laboratories station in Holmdel, New Jersey.

weather reports, he was able to attribute them to more distant thunderstorms. But even in quiet weather he heard static. This final source didn't come in bursts, but was constantly present, although it was very quiet compared to the thunderstorm bursts. As he began his study of this constant low-level static, he noticed that the source appeared to be near the sun. But if it were due to the sun, the signal would appear to be coming from the same point in the sky every 24 hours. Instead the signal repeated every 23 hours and 56 minutes. This is a time interval known as a sidereal day. The 24-hour day is the time between when the sun is, say, at its highest point one day and its highest point the next day. The sidereal day is the time it takes for a specific star to return to its highest point.

ANOTHER SOURCE OF STATIC

The timing puzzled Jansky. His measurements were making it seem as if the radio wave static he detected was being generated off the Earth. He couldn't imagine any reason for that to be true. He reported his results to the International Scientific Radio Union meeting in April 1932. He talked about the radio waves he detected from thunderstorms. Then he mentioned another source of static: a steady hiss from… somewhere. He didn't know where.

He went back to his antenna and took more data. As the year progressed the source of the signal appeared to be further from the sun. In fact, the source of the signal moved around to the nighttime side of the Earth.

By the time he had a full year of data Jansky found himself driven to an inescapable conclusion. The source of the radio signal was not anywhere on the Earth, it wasn't the sun, it wasn't even in the solar system. The source of the constant radio signal was out among the stars. Specifically, it was near the center of our Milky Way Galaxy. Quite by accident his investigation into Earthly static sources had given him the tools to measure well beyond what he had been expecting to find. He reported those results at the 1933 meeting of the International Scientific Radio Union, this time in a talk entitled, "Electrical Disturbances Apparently of Extraterrestrial Origin."

A PUZZLING NEW SIGNAL

The paper was published to great acclaim. Sort of. *The New York Times* published a front-page report on the puzzling new signal. The article even contained an assurance that the signal was not the result of "some form of intelligence striving for intra-galactic communication." That's what it wasn't, but what was it?

One would suppose scientists would be clamoring to know. But Jansky's announcement was almost ignored in the scientific community. It was the middle of the depression, and few organizations could contemplate building new apparatus and embarking upon an entirely new field of study, particularly when there was no assurance the signal represented anything of real interest. It could have been some strange internal instrumental noise or even emission due to dust within the Earth's own atmosphere. Jansky proposed to Bell Labs's management that they build a new antenna specifically to characterize

ABOVE: *Armillary sundials are devices that measure time using the position of the sun.*

The source of Karl Jansky's constant radio signal turned out to be the Milky Way.

the galactic radio signal, but since that particular static was so low it didn't affect transatlantic communications, AT&T dropped the project and assigned Jansky to other tasks.

RADIO ASTRONOMY WAS BORN

Throughout human history people have observed the light the universe sends our way. Before Jansky, scientists thought light carried the only information we could gather about the universe around us. After Jansky, scientists realized the cosmos carries energy in many different forms. Jansky died of natural causes in 1950, at the age of 44. Radio astronomy was born because of Jansky's observations, but he wasn't around to smoke a celebratory cigar. In 1973 the International Astronomical Union adopted the "jansky" as a measurement for the strength of a broad spectrum radio signal. It was a belated acknowledgement of the efforts of a young man whose desire to understand pushed him to follow his chance observation through the door to a new universe.

Little ears and big ears

Jansky's 1933 announcement almost fell on deaf ears. Several factors could have played into this. First and foremost, it was something completely new. It wasn't that scientists didn't believe that stars, galaxies, and interstellar dust could be sources of radio signals, it's just that they didn't even think to ask the question. No one understood the physics that could cause cosmic sources to emit radio waves, but that's because no one thought to ask if they could emit radio waves. Jansky, an engineer with only a Bachelor's Degree, didn't have the language or training to frame and communicate his findings in a way that astrophysicists could readily understand. So his findings weren't given the attention they deserved.

But his results weren't completely ignored.

Jansky's results caught the ear of Grote Reber, a well-to-do radio engineer living in Wheaton, Illinois. Reber was so excited by Jansky's results that he built his own radio telescopes. He mapped out the "brightness" of the Milky Way and showed that the sky was full of different radio sources. Reber reported his results, but he also didn't stimulate much interest in the scientific community. That changed at the end of World War II.

Many engineers and scientists had been trained in radar and radio during the war. They understood Jansky's and Grote's measurements. Prominent among these was John Kraus, who joined the faculty at Ohio State University. Kraus cobbled together enough money and enough help from his students to build a radio telescope he called "The Big Ear." With that, radio astronomy entered mainstream science. Now radio telescopes around the world detect everything from black holes to interstellar dust.

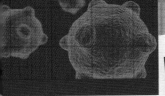

GERHARD DOMAGK

DISCOVERS THE FIRST
ANTIBACTERIAL CHEMICAL

1932

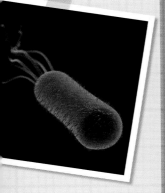

COLORING THE FIGHT AGAINST DISEASE

IN 1932 CHEMISTS, *doctors, and biologists were searching for a "magic bullet," a compound to kill infective bacteria without harming people. Bacteria are so different from people that it's reasonable to assume there must be something poisonous to bacteria yet innocuous to humans. Gerhard Domagk had been hired by the German industrial giant Interessen-Gemeinschaft Farbenindustrie AG to find that magic bullet. He found it. Although Domagk did exactly what he set out to do, without some very lucky coincidences he would have failed miserably.*

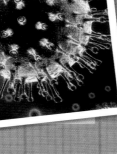

Throughout human history people have sought protection from the diseases caused by viruses and bacteria. In 1932, finally the tools were at hand.

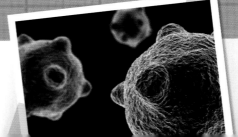

SYNTHETIC MOLECULES

Gerhard Domagk was a student at the University of Kiel when World War I began. He volunteered, was injured, and was transported back to a field hospital. He was horrified at the gruesome deaths due directly to injury. But even worse were the deaths following minor injuries. Wounded soldiers would be treated and stabilized, but then patient after patient would inexorably succumb to infection and die. After the war Domagk re-entered the University and went on to get his M.D. He was determined to find a method of fighting bacterial infections.

Interessen-Gemeinschaft Farbenindustrie, or, more simply, IG Farben, had been created when six different German synthetic-dye-producing companies merged. Germany had taken Perkin's synthetic dye discovery and created an industry by systematically modifying and efficiently producing a variety of synthetic molecules. Although the chemists were searching for dyes, they found organic chemicals with applications in fields such as photography and agriculture. IG Farben

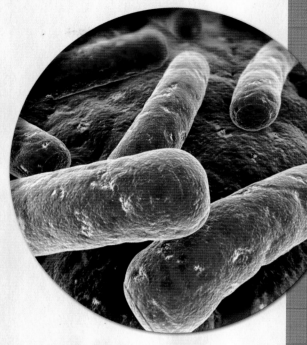

Bacteria spread in hospitals during World War I and people with only minor injuries would succumb to infection and die.

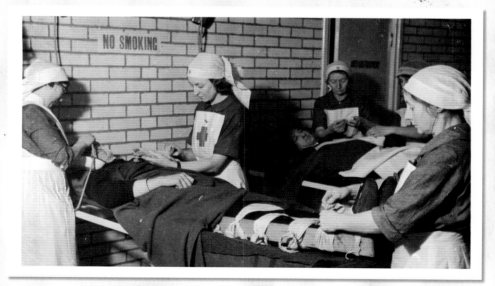

British Home Service nurses during World War II attend to an injured man, anesthetizing him while they tie his legs up in splints. By this time, thanks to Gerhard Domagk, people had more knowledge about bacteria and the importance of warding off and fighting bacterial infection.

officials were also hoping to find pharmaceutical applications. They had little evidence to support that hope, but that didn't keep them from hiring Gerhard Domagk to investigate the potential disease-fighting action of their chemicals.

PRONTOSIL RED

In 1932 Domagk was the Director of IG Farben's Laboratory for Experimental Pathology and Bacteriology. His job was to test every chemical for antibacterial activity. IG Farben's synthetic chemists had just developed a new orange-red wool dye. They had been disappointed in the color-fastness of an existing class of dyes; so they resurrected a technique from earlier generations of dyes. They made an "azo" form of the dye, which means they connected the dye molecule through a nitrogen atom to another molecule, in this case a molecule called a sulfanilamide. The new dye, named Prontosil Red, was fast to wool, that is, it didn't wash out easily. Because it bound tightly to the proteins in wool, the chemists reasoned that it would bind equally tightly to protein-laden bacteria.

Domagk tested the effect of the dye on a bacterial culture. It didn't do anything. Domagk figured he didn't have anything to lose by testing in mice. He infected mice with a particularly virulent strain of *Streptococcus pyogenes*, one of a class of bacteria responsible for pneumonia, rheumatic fever, childbed fever, and other potentially lethal infections. He injected the dye into the infected mice and they survived. The results were so dramatic he immediately planned for clinical trials on human patients.

There are two stories about what happened next; probably both are true.

Before the clinical trials could take place, an infant at the local hospital developed a certainly fatal streptococcal infection. Dr. Richard Foerster had no way to save the infant, but he had heard rumors of Domagk's early success. Foerster requested some of the compound, injected it and cured the child. At around the same time Domagk's own daughter had fallen and put a knitting needle through her hand. The wound became infected, again with what was known to be fatal *Streptococcus*. He injected his daughter with Prontosil Red and she recovered. Although those patients demonstrated the effectiveness of the drug, its most famous early use came when President Franklin D. Roosevelt and his wife Eleanor asked their doctor to administer Prontosil Red to their sick son in 1936. Franklin Jr. was cured and the drug became sought-after in the United States.

Domagk had found what he had set out to find. Prontosil Red worked. But for all the wrong reasons. Domagk and the IG Farben chemists had it all wrong. The success of the new drug was due to chance as much as it was due to diligent preparation.

Their first wrong assumption was revealed by the husband and wife scientific team of Jacques and Thérèse Trefouel, who discovered that the body separates Prontosil Red into the dye part and the sulfanilamide. Without the biochemical reaction in the body separating the two parts of the molecule, it does nothing. The only reason the sulfanilamide had been put there was to improve the dye, but the sulfanilamide is the active part of the drug and the dye isn't necessary at all! Once that was understood, a whole series of sulfanilamide-containing "sulfa drugs" were synthesized—meaning IG Farben never realized significant profits from Prontosil.

The most famous early use of Prontosil is when President Roosevelt and his wife Eleanor had it administered to their sick son. After Franklin Jr.'s cure, the drug became sought-after in the US.

Another wrong assumption: the sulfanilamide doesn't work by grabbing onto proteins and paralyzing the bacteria; it works by not allowing the bacteria to multiply. Bacteria need to make a compound called folic acid to copy their DNA and reproduce. The sulfanilamide stops the bacteria from making folic acid; so the infection can't progress. The drug doesn't harm humans because we don't make our own folic acid, we get it from our food.

Domagk made assumptions about why Prontosil Red was a good candidate as a disease-fighting compound. His assumptions were wrong. He was lucky, but he was also good. His thorough testing even in the face of unpromising preliminary results brought the first broadly effective antibiotic to humanity. He was acknowledged with the Nobel Prize for Physiology or Medicine in 1939, but the Nazi regime forced him to refuse the award. He finally was able to accept in 1947, although the monetary award had been returned to the prize pool. Before Domagk's work, chemists hoped they might eventually find some chemical antibiotic. After Domagk, scientists knew it could be done. Largely because of his lead, since 1935 hundreds of antibiotic compounds have been identified, saving millions of lives.

The use of antibiotic compounds such as sulfa has become commonplace, and they are administered to the wounded on battlefields by fellow soldiers.

Guests at the seventh meeting of Nobel Prize-winners, Lindau, Germany, 1957. On the left is German pathologist Gerhard Domagk (1895–1964).

Cashing in and cashing out on the discovery

In 1937 sulfa drugs were in demand around the world. Companies wanted to cash in on the sulfanilamide "magic bullet" craze. In the United States at the time, if you wanted to offer a drug to the public all you needed to do was put something in a package and sell it. That's just what the S.E. Massengill Company did. They hired chemist Harold Watkins to make his own "Elixir of Sulfanilamide." Watkins prepared the drug in a ten percent solution of diethylene glycol mixed with raspberry flavor and saccharin. Diethylene glycol is a sweet, syrupy compound. What Watkins didn't know: diethylene glycol is also toxic. The elixir killed 105 people in two months. When he learned of the deaths, Watkins killed himself. The Massengill Company was fined $26,000 for false labeling. But the case led to recognition of the need for safety regulations. In the United States federal safety laws were passed: future drugs would need to be tested for safety. In a way, sulfa drugs saved even more lives indirectly than they did directly.

ROY PLUNKETT

INVENTS DUPONT™ TEFLON®

1938

A BIG SLIP-UP

REFRIGERATORS and air conditioners work using the cooling created as a compressed gas or liquid expands. Although a practical refrigerator had been invented in the mid-1800s, the cooling gases were toxic, and accidental leaks killed hundreds of people. When General Motors developed a new refrigerant, dichlorodifluoromethane, they partnered with Frigidaire and DuPont to further develop and market the product. In 1931 DuPont trademarked the new refrigerant, a chemical in a group known as chlorofluorocarbons (CFCs), under the name of DuPont ™ Freon®, but the company still was actively researching potential replacement compounds and processes.

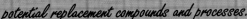

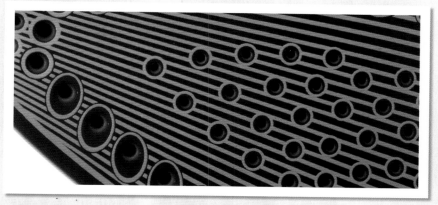

DuPont™ Teflon® is used for irons as well as cooking utensils and cables because it is non-stick, light, non-reactive, and very heat resistant.

TETRAFLUOROETHYLENE (TFE)

DuPont hired a fresh Ph.D. student, Roy Plunkett, and set him to the task of investigating chlorofluorocarbons for new refrigerants. Plunkett failed miserably. In attempting to make a new gas, he ended up making a new solid. Because he didn't discard his failure, he developed a product with worldwide applications.

In 1936 Roy Plunkett got his doctorate in chemistry from Ohio State University and went to work at DuPont's Jackson Laboratory at Deepwater Point, New Jersey. In April of 1938 he was working to develop new CFC refrigerants. On April 5, 1938, Plunkett finished producing 100 pounds of a gas he needed as one of the components for the CFCs he was trying to make. The gas was tetrafluoroethylene. Tetrafluoroethylene is a molecule with two carbon atoms tightly attached to each other through what is called a "double bond," with two fluorine atoms attached to each carbon atom. Plunkett made the gas and stored it in some small gas bottles, cylindrical metal containers that can hold pressurized gas. Plunkett had completed making the tetrafluoroethylene he needed. But it was late in the day; so he took the several small cylinders of gas he had made and put them "on ice." Actually, he put them on dry ice, solid carbon dioxide with a temperature of about -167°F (-75°C). He came in the next day, and he and his assistant Jack Rebok connected one of the tanks to the reaction apparatus. Nothing came out.

Plunkett figured the container must have leaked overnight. But he weighed the container and it was the correct weight. Then he supposed the valve was broken. But he checked the valve and it appeared to be fine. So Plunkett and Rebok pulled another filled bottle off the pile and tested that one. The same thing happened. Now their curiosity was piqued. The gas they were looking for wasn't there, but something was in these bottles. They removed the valve from one of the bottles and gave it a good shake. Some flakes of a white solid fell out. Finally their curiosity drove them to take a bottle and saw it in half.

When working with TFE, oxygen must be scrupulously excluded, as it catalyzes autopolymerization, which can take place with great force.

SPONTANEOUS POLYMERIZATION

They found the inside to be coated with a white, waxy solid. Plunkett realized the tetrafluoroethylene must have spontaneously polymerized. That is, the individual gas molecules must have connected up with each other to make large molecules. This was completely contrary to the understanding of the day, which held that molecules of this fluorine-carbon type couldn't form polymers. Obviously, this was not going to help him do his job: make a gaseous or liquid refrigerant. It was a dead end.

But Plunkett didn't stop.

Perhaps he was intrigued by the "impossibility" of the reaction that occurred. Perhaps he was driven by a desire to understand what he observed. Instead of tossing out the ruined cylinders, Plunkett spent the next several days characterizing the new material. He found some unique properties.

First, the material was slippery. Really slippery. Second, the material was resistant to acids, oils, alkalis, solvents—hardly any chemical would affect it! Third, the compound was heat resistant, maintaining its properties up to 500°F (260°C) and beyond.

SPACESUITS TO MUFFIN TINS

After more study, Plunkett identified the new material as polytetrafluoroethylene, or PTFE. DuPont trademarked the material as Teflon®. PTFE has all those amazing properties because of its chemical structure. Magnified, it would look like a bunch of carbon atoms all hooked together in a long string. Then, all around that string would be fluorine atoms, kind of like a protective sleeve around the carbon string. Many compounds that damage organic molecules work by attacking the carbon bonds one way or another, but in PTFE the carbon bonds are all protected. And fluorine binds very tightly to carbon, which means it doesn't interact much with other molecules. Because bound fluorine is so "unfriendly" to contacting other molecules, PTFE is slippery.

Teflon® has found its way into everything from spacesuits to muffin tins, all because of the chance observation of a prepared individual whose desire to know outweighed his need to meet a schedule.

DuPont chemist Roy Plunkett was investigating CFC refrigerants in 1938. He was trying to make 100 pounds of an intermediate compound, tetrafluoroethylene, when the gas polymerized and Plunkett discovered he had made Teflon®.

DuPont™ Teflon® is incredibly slippery because its components do not interact with other molecules.

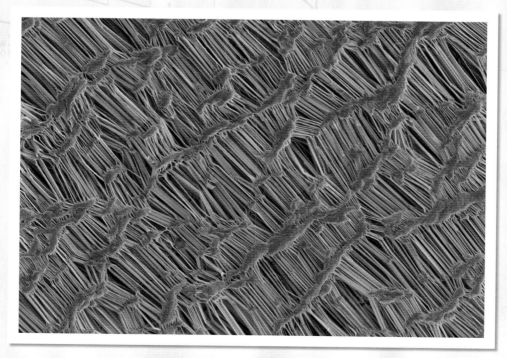

DuPont™ Teflon®, as seen through a scanning electron microscope (with artificial color shading).

DuPont™ Teflon® was first employed by the US Army for the Manhattan Project, where it was applied in the manufacturing of containers, pipes, and valves that would carry unranium hexafluoride (an intermediate compound used to concentrate radioactive nuclear material). The Manhattan Project later developed the atomic bomb. DuPont™ Teflon® is now sold commercially worldwide for various household and industrial applications.

Exploding into the marketplace

It didn't take long for Plunkett and his DuPont colleagues to realize they had a compound with unique—and valuable—properties. But were those properties valuable enough? PTFE was expensive to produce. Without some significant effort put into optimizing fabrication, could it be produced at an affordable price? Luckily for generations of cooks, to DuPont's first PTFE customer, money was no object.

During World War II the United States' "Manhattan Project" was designed to evaluate the potential explosive capability of radioactive nuclear material. The project was incredibly complex, requiring technological and scientific advances in several disciplines. For example, just obtaining the fuel was a chemically challenging endeavor. One reason for the challenge was that a necessary intermediate product—uranium hexafluoride—was an incredibly corrosive gas, and would rapidly eat away at containers, pipes, and valves with which it came

in contact. General Groves, the US Army officer in charge of the project, heard from some DuPont acquaintances about the new chemically resistant material. He didn't care about cost; so PTFE found its first customer. It wasn't until 1960 that the PTFE production process became streamlined enough to affordably offer non-stick DuPont ™ Teflon® cookware to the general public.

CHAPTER 4

COMPLEX CONNECTIONS: SERENDIPITY ACROSS DISCIPLINES

★

A WARM POCKET LEADS TO A HOT COOKING TECHNIQUE
Percy Spencer invents the microwave oven, 1945

ONE DROP DOES IT
Harry Coover invents cyanoacrylate Superglue, 1951

A SLICK INVENTION
Patsy Sherman invents Scotchgard™ water repellent, 1953

PIECING IT TOGETHER
Francis Crick and James Watson decipher the structure of DNA, 1953

The trend in science is like the trend in any discipline: the more that is already known, the harder it is to move into territory no one else has discovered. Hand-in-hand with this is ever-increasing specialization. If a scientist is going to do something new in a field, they first need to understand what has already gone before. But they also have to fit their quest for knowledge into a human lifetime.

Before 1800, natural philosophers trying to understand the world could contemplate, well, everything. Even granting that Aristotle was probably the greatest thinker in the history of the world, the scope of the problems he considered is representative of the range of topics accessible to natural philosophers for thousands of years. Anyone who dedicated a fair percentage of their life to study could write about everything from politics to biology to physics and be just as authoritative as anyone in the world. But in the 1800s people began to devote their lives to "science," and anyone wanting to speak authoritatively in the field needed to be similarly dedicated. In the era from 1900 to 1940 scientists became even more specialized, devoting their lives to "chemistry" or "physics," and again, similar dedication was needed to become an authority.

After 1940 the specialization became even more intense. In this era, it became reasonable to be called not just a scientist, not just a chemist, but an organometallic chemist. It was a necessary specialization because, to work at the forefront of a field, expanding knowledge, a scientist needed to be prepared enough to recognize when a new observation was significant.

But the world doesn't always divide itself into the clever categories that we humans devise. Although depth of knowledge was important, breadth of understanding was also critical. One way in which scientists balanced depth and breadth was to expand their teams. All of the accidental discoveries presented in this section required the participation of more than one person. In most of these cases, each of the several people involved made significant contributions to the discoveries.

This had two implications for accidental discovery. First, at least one person needed to be prepared enough to be confident that the accidental discovery was worth pursuing. Second, the person who felt that a chance observation was significant needed to convince other members of the team of that importance. In this era there were as many opportunities as at other times. But for those chance opportunities to turn into accidental discoveries, preparation and desire were more important than ever.

KEEPING THE BEAT
Wilson Greatbatch invents the implantable pacemaker, 1958

DON'T GUM UP THE WORKS!
Stephanie Kwolek invents DuPont™ Kevlar®, 1965

RECEIVING THE LONG-DISTANCE CALL
S. Jocelyn Bell discovers pulsars, 1967

PLASTICS GET CURRENT
Alan MacDiarmid, Hideki Shirakawa, and Alan Heeger invent intrinsically conducting polymer, 1977

A WARM POCKET LEADS TO A HOT COOKING TECHNIQUE

★

EVERY DAY MICROWAVE OVENS *heat millions of cups of coffee and tea, countless bowls of soup and popcorn, and too many leftovers and ready-prepared meals. The efficiency of the microwave oven has made it practically a necessity in the modern kitchen. The whole development was triggered by a chance occurrence: a melting chocolate bar in an industrial laboratory.*

ELECTROMAGNETIC WAVES

In 1945, 51-year-old Percy Spencer was working in Raytheon's laboratory in Lexington, Massachusetts. He was trying to improve the efficiency of the magnetron, a device used for generating microwave radiation. During World War II, magnetrons had become critical elements of radar technology. Radar works by sending an electromagnetic wave into an area, then detecting reflections from the objects it strikes. Initially, radio waves were used, but their long wavelength, typically in the range of 330 ft (100 m) or so, meant that smaller objects, such as 33-ft (10-m) wide aircraft, did not reflect them very efficiently. In addition, detecting the small returned portion of these long wavelengths required a big antenna. Microwaves are related to radio waves, but they are much shorter, on the order of a fraction of an inch; so they're more efficient and require much smaller antennas.

From laboratory to the home

Although the essential principles of the microwave oven were all demonstrated in Spencer's initial experiments, it took several decades for the microwave oven to reach the status of domestic necessity. The first commercially available "Radarange" (the winning name in an employee contest) was nearly 7 ft (2 m) tall and weighed 750 lb (340 kg). Also, because the magnetrons were water-cooled, it required its own plumbing hookup as well. As might be imagined, sales were not too brisk.

It wasn't until Raytheon acquired the Amana Refrigeration Company (nearly two decades later!) that the necessary expertise—and corporate priority—was mustered to produce a commercially viable home unit. At $495, 1967's countertop Radarange began to be within reach of many households. Now inexpensive small models pepper college dormitories around the world, and sales of microwave ovens exceed sales of "traditional" ovens in many countries. All from a melted chocolate bar!

Built in 1947, the Radarange was the first microwave oven ever made.

Melting chocolate in an industrial laboratory was the accidental spark behind the invention of the microwave oven.

AN INNATE UNDERSTANDING

During the war, Spencer had made a series of design refinements to ease the manufacturability of magnetron tubes. His design modifications skyrocketed production from 17 a week to 2,600 a day. So Spencer had demonstrated an innate understanding of magnetron science and technology, which is why he, a man who had never finished high school, was in charge of a development laboratory at an advanced research and development facility. On one particular day, he was also hungry, and that's why he changed the world.

Spencer had a chocolate bar in his pocket as he wandered through the magnetron laboratory. He felt a strange sensation, and noticed that the candy had started to melt. His first thought: the microwave radiation from the magnetron had heated the chocolate. To check his idea, he began a series of tests. In an eerily prophetic move his first test object was an ear of Indian corn, which (as millions of teenagers can now happily affirm) popped! He grabbed a colleague and a raw egg and repeated his experiment—with the egg, not the colleague. In front of the magnetron the egg started to twitch and rock as it internally heated. Spencer's colleague leaned in to get a closer look just as the egg exploded, leaving the colleague spattered, and Spencer with an idea.

LEFT: *Now eaten all over the world, popcorn was the first object to be deliberately tested by Percy Spencer using radiation from the magnetron.*

FACING PAGE: *Woman demonstes the stove of the future, in which high frequency waves cook right through the food. The invention was a forerunner of the microwave oven, commonly in use today.* Illustrated London News, 1947.

PREPARATION AND OPPORTUNITY

Others had reportedly observed the heating effect of microwave radiation, but Spencer had not just the preparation and opportunity, but also the desire to follow his chance observation to its logical conclusion. He and his colleague P. R. "Roly" Hanson embarked on a development program they called "The Speedie Weenie" project. They constructed a metal enclosure and sent microwaves into it. The microwaves reflected off the metal and concentrated in a small area, zapped the hot dog, and cooked it in moments. Although it took decades of development to produce economically viable microwave ovens, the essential design elements were contained in Spencer's initial set of experiments. Microwave ovens are now all equipped with beeps, buzzers, and buttons, giving them the capability to prepare complicated meals. But perhaps things haven't changed all that much: microwave ovens are commonly used to melt chocolate and pop popcorn.

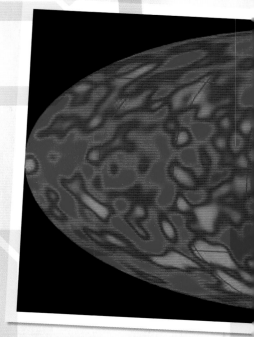

Whole sky image of the cosmic microwave background made by the MAP (Microwave Anisotropy Probe) spacecraft. The colors show variations in the temperature of the universe in all directions, as revealed through measurements of microwave radiation.

Microwave ovens being manufactured in a factory assembly line.

Microwaves make pre-prepared meals ready to eat in a matter of minutes.

Microwave ovens are now commonplace in kitchens all over the world, but the discovery that microwave radiation could heat food was an accidental one.

THE ORIGINAL SUPER GLUE CORPORATION

HARRY COOVER

INVENTS CYANOACRYLATE SUPERGLUES

1951

ONE DROP DOES IT

IT'S A MEMORABLE IMAGE. *A convertible filled with people suspended from a pair of metal plates hanging from a crane. The metal plates could have been welded to each other or bolted together, but they weren't. They were held with a few drops of cyanoacrylate adhesive, popularly known as Superglue. There were other dramatic demonstrations of the quick-acting adhesive as well, but none of them would have been possible without an accidental discovery. Well, actually an accidental discovery made twice.*

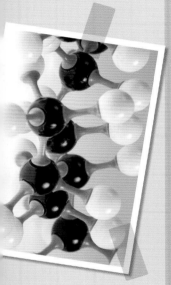

Superglue gets its adhesive properties from the molecular structure of its component parts.

GETTING POLYMERIZE-READY

While pursuing his doctorate at Cornell University, chemist Harry Coover worked with the Kodak Company. His job, like so many others in 1942, was connected with the war effort. He was tasked with developing an optically clear plastic that could be used in molded precision gunsights. One compound he tried was a class of molecules called cyanoacrylates. Coover was interested because they connect with each other easily, that is, they polymerize readily. But the cyanoacrylates he made were a pain in the neck. They stuck to everything. At the time, Coover just wanted the cyanoacrylates to go away so he could do his job. So he recorded his results, put the cyanoacrylates away, and got back to work.

Years later, in 1951, he was working at Kodak's Tennessee Eastman Company in Kingsport, Tennessee. Now he was in charge of an effort to find a stronger, tougher, more heat-resistant material for jet airplane canopies. Existing canopies were made from polyacrylates, such as poly(methyl methylacrylate) (PMMA), perhaps better known by one of its several trade names such as Plexiglas or Lucite. So Coover's team was looking for other possible acrylates. One of Coover's assistants, Fred Joyner, was evaluating possible candidate materials from members of the acrylate family. One of those candidates was cyanoacrylate.

THE INDEX OF REFRACTION

Joyner was thorough. He set out to measure all the properties of the material. One of the most critical properties for a transparent optical material is called the index of refraction. The index is a measure of how quickly light travels through a material, which relates to how much a window or canopy of the material will distort the scene. Cyanoacrylate is normally a liquid; so Joyner had to use an expensive, specialized apparatus to make the measurement. He put the cyanoacrylate

Superglue is now commonplace in most households.

liquid between the two prisms in the apparatus and he made his measurement. But when he tried to take the two prisms apart, he couldn't. The cyanoacrylate had glued the two pieces of glass together. Joyner had to confess his $700 mistake to his boss, Coover. It couldn't have been easy, because in 1951 $700 was quite a large amount.

To Joyner's delight, Coover was not upset. When Coover heard the story, he recalled his own initial encounter with troublesome cyanoacrylate, the compound that stuck to everything. This time, however, Coover wasn't feeling anything like the pressure of wartime that he had felt in 1942. He was open to the possibilities the compound presented. The same afternoon that Joyner reported the ruined prisms, Coover grabbed some of the liquid and glued everything he could lay his hands on: glass, rubber, metal, wood, paper, plastic—nothing was safe that afternoon. At the end of the afternoon, cyanoacrylate adhesives were born.

EASTMAN 910

Kodak named the compound "Eastman 910," because it bonded by the time a user counted "1, 2, 3, 4, 5, 6, 7, 8, 9, 10." The material bonded so rapidly because water molecules would trigger a cascade of polymerizations, where one cyanoacrylate molecule would attach to another. It didn't take much water, just a few molecules—an invisible amount—would precipitate the reaction.

The bond was strong also. The detailed explanation is rather technical, but the short explanation is that when two cyanoacrylates joined together an electron moved all the way from one side of one molecule to the opposite end of the second molecule. To break the bond, the electron needed to be moved all the way back, and that didn't happen very easily. And, that's just the start. Each time a new cyanoacrylate monomer joins onto the polymer, the electron gets pushed even further from its home, making the bond even stronger.

The original Superglue from the Superglue Corporation in California.

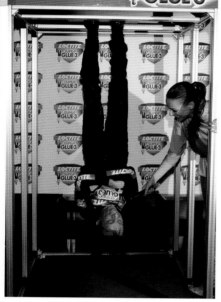

Alain Robert, also known as the "French Spiderman," hangs upside down to test the new Loctite Super Glue-3 in 2008.

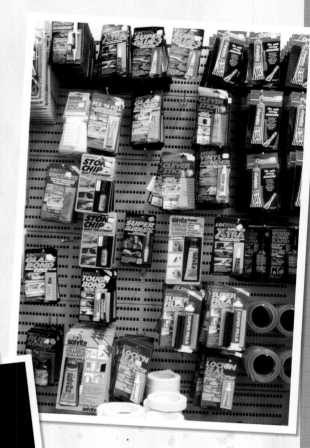

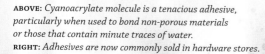

ABOVE: Cyanoacrylate molecule is a tenacious adhesive, particularly when used to bond non-porous materials or those that contain minute traces of water.
RIGHT: Adhesives are now commonly sold in hardware stores.

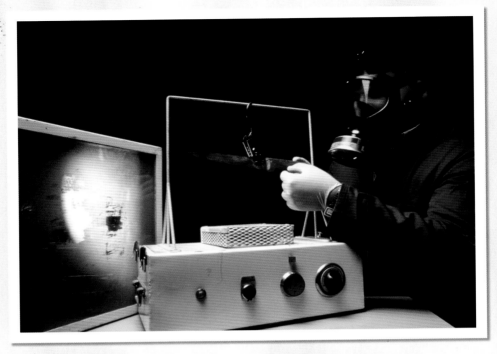

Cyanoacrylate is used in fingerprint detection to reveal fingerprints on murder weapons.

Coover identified the two chemical characteristics as good marketing gimmicks as well. He discovered the new glue, now he became its most fervent champion. He even appeared on the television show, "I've got a Secret," where he provided an effective demonstration of "Superglue." He bonded two surfaces with a single drop of Superglue then, moments later, Coover and program host Garry Moore hung from a bar suspended through the single drop.

STICKING AROUND

There are several different cyanoacrylates, and just about all forms have been made into some kind of fast-setting, strong-bonding adhesive. There are many different formulations marketed under a number of trade names, but all are descended from Coover's initial discovery. Cyanoacrylate adhesives have been used to join fragmented bones at archeological sites, to assemble parts for satellites, and enhance fingerprint identification at crime scenes. Just like the pesky compound when it first appeared in Coover's lab, cyanoacrylate adhesives are sticking around.

LEFT: *Cyanoacrylate is used in tissue repair and for various originally unintended purposes.*

FACING PAGE: *Coover recognized the potential for cyanoacrylate to be used in different situations, and came up with a spray that was used on US war casualties to stop bleeding until proper treatment could be administered by medical professionals.*

A "killer app" that saves lives

In 1942, Coover had not been open to the opportunity presented by his first chance observation. When chance stepped in a second time, he was more than ready. He recognized, developed, and publicized cyanoacrylate superglue as a general adhesive, but he also recognized another application: as a medical adhesive.

Coover recognized that his glue could be used to promote tissue mending without sutures or other mechanical intervention. He also developed a cyanoacrylate adhesive spray applicator that was used to treat US casualties in the Vietnam War. The glue, a variation on the standard superglue compound, would be sprayed on open wounds where it would rapidly stop the bleeding and stabilize the patient for later treatment. Be careful doing this

at home, though. Medical cyanoacrylates are polymers made up of longer molecules than the standard adhesive cyanoacrylates, and the shorter-molecule versions aren't as well-suited for medical applications.

Today cyanoacrylate medical adhesives are used to promote healing by attaching organs, structures, or tissues to each other. Not bad for an accidental discovery that had to be made twice!

PATSY SHERMAN

INVENTS SCOTCHGARD™
WATER REPELLENT

1953

A SLICK INVENTION

THE STORY OF PATSY SHERMAN *and her colleague Sam Smith illustrates one class of accidental discovery, especially common in chemistry. Chemists are very rarely told to go play in the lab. Just as in every other profession, chemists have a job to do. Often that job involves making a compound for a specific use. It's almost certain that when chemists set out to make a compound, they'll be successful. That is, they won't necessarily make a molecule with the characteristics they're looking for, but they'll make something. If it's not what they're looking for, will they be open to new possibilities? Patsy Sherman was.*

Reformulated Scotchgard™ is now used all over the world to protect fabric, furniture, and carpets from stains.

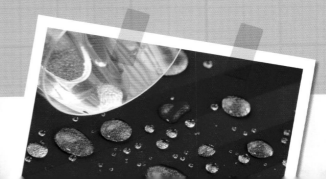

AN INTERMEDIATE STEP

Patsy Sherman was born in 1930, graduated from Gustavus Adolphus College in Saint Peter, Minnesota in 1952, and was immediately hired by the Minnesota Mining and Manufacturing Company (3M). She began work as a chemical researcher. Sherman was one of the few women in the field, and 3M was one of the few companies open to hiring women for the technical staff. She was assigned, with her colleague Samuel Smith, the task of creating a synthetic rubber resistant to degradation from oil and fuel. The intention was to develop a new material for jet aircraft fuel hoses. A secondary goal—maybe the primary one for 3M—was to expand the market for fluorochemicals, an area where the company had significant technical expertise. Sherman and Smith never achieved their primary goal. But a laboratory accident got them on track to reach their secondary goal.

Scotchgard™ is a particularly efficient cleaner and degreaser of hydraulic pipes used in the aviation industry.

Sherman had made a fluorochemical-latex emulsion as an intermediate step in her search. One day in 1953, a laboratory assistant dropped a flask containing the emulsion and it splashed all over the place, including on the assistant's shoes. As anyone would do, they cleaned up the mess. But, try as they might, they couldn't get the compound off the assistant's white canvas shoes. The color of the canvas was unchanged, but the spill couldn't be cleaned up. In fact, it couldn't even be wet. Where the compound had spilled, water just beaded on the surface of the canvas shoe.

This behavior was unexpected enough to draw Sherman's interest. She and Smith studied the canvas shoe. The area spilled upon was impervious to solvents, and repelled liquids of all sorts. The stain didn't just resist liquids, it also resisted dirt. This was something worth working on!

Sherman and Smith spent the next few years investigating, experimenting, and developing the chemical. Although it must have been extremely gratifying for Sherman to observe her accidental discovery developed, there were many frustrating moments as well. When her compound was tested inside the textile mill, she had to wait outside, as women were not allowed to enter. Sherman and Smith had developed Perfluorooctane Sulfonate (PFOS), a type of compound called a "block and graft copolymer containing water-solvatable polar groups and fluoroaliphatic groups." In 1956, 3M put it on the market as a fabric stain repellent. With such a melodious title, it's easy to see why 3M selected a slightly different name for their product: Scotchgard™. Sherman and Smith continued to work on this type of fluorochemical, developing applications for cleaning car upholstery, sealing carpets, and protecting photographic prints.

Sherman attributed her great success to her willingness to always keep an open mind and never overlook the possible significance of an apparent failure. Speaking at the celebration of the US Patent Office's 200th anniversary, she said, "you can teach yourself not to ignore the unanticipated." She noted that many inventions have been the result of just noticing something no one conceived of before. She made the most of her chance opportunity, creating a product whose annual sales have been as high as $300 million.

Unfortunate coincidence

In May 2000, 3M announced they were reformulating Scotchgard™ to replace Perfluorooctane Sulfonate (PFOS). PFOS, a synthetic molecule that was the major constituent of Scotchgard™, had been found in environments, animals, and people around the world. It's a persistent molecule, meaning it doesn't break down very easily. That was initially regarded as a desirable feature, after all, it doesn't do much good to spray on a fabric protector that will degrade in a couple weeks and no longer be a fabric protector. But concerns over possible health risks prompted 3M to reformulate Scotchgard™ with more benign chemicals. PFOS was designed to reduce waste by extending the usable life of a variety of products. The unhappy coincidence was that it turns out to have possible health impacts on animals and humans.

A more unfortunate coincidence involved chlorofluorocarbons, or CFCs. CFCs were miracle chemicals when they were developed: millions of lives were saved either directly or indirectly by the inexpensive, safe refrigeration enabled with CFCs. Then, during World War II, it was discovered that CFCs made excellent propellants for spray cans. CFCs are very chemically stable. Again, this is an advantage, because a refrigerator is supposed to refrigerate and a spray can is supposed to spray. They wouldn't work if the CFCs degraded.

CFCs used as propellants were expelled into the air. Here the unfortunate coincidences begin. It turns out that CFCs are very stable, unless they're exposed to strong ultraviolet light. The first coincidence is that expelled CFCs are light enough that they float high into the atmosphere where the ultraviolet light from the sun is much stronger—— strong enough to split the molecules apart. The second coincidence: when CFCs break apart they release a chlorine atom, and chlorine atoms act as a catalyst for breaking apart ozone molecules. The third coincidence is that CFCs are just light enough to hang out at the same height in the atmosphere where the bulk of the Earth's protective ozone stays. If CFCs were just a little lighter, they'd be above the ozone layer. If they were a little heavier, they'd be beneath the ozone, where they'd be protected from most of the UV radiation. And if they didn't release a chlorine atom, then they wouldn't present a threat to atmospheric ozone. This series of unfortunate coincidences is a major factor influencing global climate change.

FRANCIS CRICK AND
JAMES WATSON

DECIPHER THE
STRUCTURE OF DNA

1953

PIECING IT TOGETHER

DETERMINING THE STRUCTURE *of DNA is one of the most celebrated scientific discoveries of the last century. The "double helix" structure is now almost universally recognizable. Nearly all accidental discoveries would eventually be made by someone, but that's even more true for this discovery. Several research groups were on track to reveal the structure of DNA. Someone was going to do it and chance played a role not insofar as the discovery being made, but as far as who made it. James Watson and Francis Crick succeeded, but without a particular convergence of geographic proximity and interpersonal relations the credit would have gone elsewhere.*

DNA is often compared to a recipe or code because it contains the instructions needed to make components of cells.

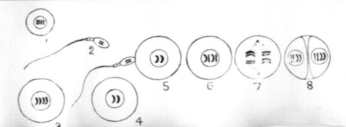

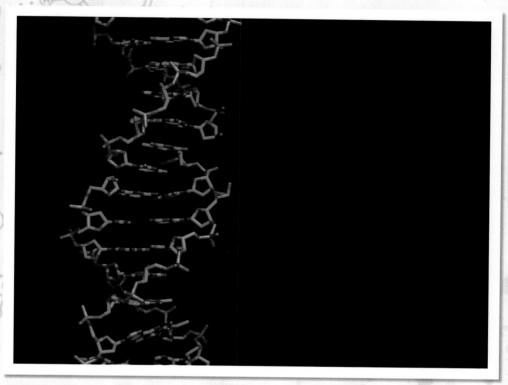

DNA strands are made up of alternating phosphate and sugar residues which hold a series of purine and pyrimidine bases.

BIOLOGICAL SECRETS

The first decades of the twentieth century revealed a number of biological secrets, but one huge question remained: how was the pattern of human life transmitted? The question had two parts. First, how did children get traits from their parents? Second, how did each cell in an organism know how to develop?

The first question can be summarized simply. When two reproductive cells come together (usually a sperm and an egg) they merge into a single cell called a zygote. Looking at their zygotes, it is nearly impossible to distinguish a chicken from a raccoon from a human, yet somehow each zygote grows into exactly the kind of organism it's supposed to be. How does that happen? How does information get transferred from parents to offspring?

The second question is a bit more complicated but it's essentially the same. As an embryo grows from the zygote, the cells begin to differentiate. That is, they turn into different types of cells. Where is the information that directs, say, a liver cell to grow into a liver cell, and then tells it how to do it? How does information get transferred from the single zygote to the billions of cells that make up our bodies?

Somewhere within the cells is something that contains all that information. But where?

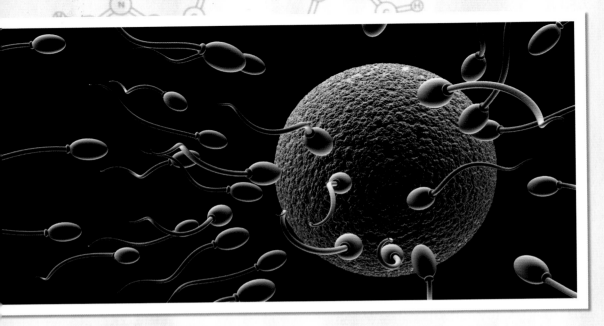

Sperm fertilizes an egg to form a zygote.

A VERY COMPLEX MOLECULE

Scientists figured whatever that "something" was would need to be a very complex molecule, because it had to carry all that complex information, but, at the same time, it had to be nearly mistake-free. Put another way, not too many people are born as fish, or have a puppy tail growing from their cheek. So biologists had looked at the most complex biological molecules they knew: proteins. Proteins are long and complex and have very complicated structures that create strings and tubes and plane surfaces all mixed together. Proteins could certainly carry the information needed to guide the growth of living organisms. So it came as a surprise to many when, in 1943, it looked as if another molecule was responsible.

AN ELEGANT EXPERIMENT

In 1943 Maclyn McCarty, Oswald Avery, and Colin McLeod had completed an elegant experiment where they allowed DNA from one strain of *Pneumococcus* bacteria to be incorporated into another strain. The bacteria didn't grow to be the same type as their parent cell, they grew to be the same type as the cell that supplied the DNA. It was a simple, yet conclusive experiment, but many within the scientific hierarchy could not accept it. DNA was a long polymer built out of a simple sugar, phosphate, and four other small organic molecules. The four other molecules consisted of bases known as purines—adenine and guanine (A and G)—and the pyrimidines, thymine and cytosine (T and C). Pyrimidines are molecules built around a 6-atom ring of carbon and nitrogen, while purines are built around one 6-atom ring combined with one 5-atom ring. To one of the leading biologists of the day, Alfred Mirsky, DNA was far too simple to carry all that complex information, and anyone who believed it did must be simpleminded. Mirsky

discredited the idea at every opportunity (which may have denied McCarty, Avery, and McLeod the Nobel Prize they probably deserved).

Even with Mirsky's opposition, other biologists were convinced by the McCarty-Avery-McLeod experiment. Rather than spend their time arguing that DNA couldn't carry all that information, they tried to figure out how it could. Chemist Erwin Chargaff determined that DNA molecules had as many purines as pyrimidines. That is, they had as many A and G molecules as they had C and T molecules. It was an interesting observation, But it didn't answer the question of how the chemically simple DNA macromolecule could execute such a complex biological process. To do that, scientists needed to understand the structure of DNA, how it was put together.

But it wasn't an easy task. Living organisms are not like test tubes containing separate chemicals. In living beings different molecules are mixed in a jumbled, sticky mess. DNA can be separated out, but it's a globby conglomeration of goop. Chargaff had already taken the step of breaking the DNA molecules apart and measuring the pieces, but now the hard job was figuring out how those pieces went together.

At CalTech Linus Pauling had just recently described the structure of proteins, an insight he arrived at largely by building scale models. Pauling and his group were now trying to figure out the structure of DNA. John T. Randall headed the Biophysics Unit at King's College in London, where physicist Maurice Wilkins led the effort to identify the structure. A third group was almost accidentally in the mix.

BRITISH SCIENTIFIC TRADITION

Sir Lawrence Bragg led the Cavendish Laboratory at Cambridge University where Francis Crick was working toward his doctorate in physics. A US zoologist, James Watson, had come over to work at the Cavendish. The British scientific tradition was to avoid direct competition with other British scientists. That tradition was, literally, foreign to Watson, who convinced Crick that they could work on the problem even though it put them in direct competition with Wilkins. Crick and Watson spent long hours bouncing ideas off each other. They also, like Pauling, spent endless hours playing with models of the molecules that compose DNA. But Watson and Crick weren't gathering any data or running any experiments; so they didn't really make any progress.

Either Pauling, or Wilkins, or Watson and Crick were going to win the race to determine the structure of DNA. As much as all three groups were working toward that goal, at this point their hard work would mean less than a series of random coincidental circumstances. The story of one of the most celebrated scientific discoveries of all time was about to turn into a soap opera.

Photo 51 is an X-ray diffraction photograph of DNA taken by Rosalind Franklin and published in 1953. The image results from a beam of X-rays being scattered onto a photographic plate by the DNA.

Wilkins, at King's College, knew this problem required him to use the most advanced tools at his disposal. A technique called X-ray diffraction was just what he needed. When X-rays bounce off a molecule, the direction they bounce is determined by the orientation and the spacing of the atoms within the molecule. So Wilkins taught himself about X-rays, modified an X-ray system to focus down small enough to get good information about DNA, and then extracted a single strand of DNA from a tangled mass. The data he got hinted at a possible helical structure, but the signal from just one molecule is very small; so he couldn't be sure.

In order to get a good signal from X-ray data, the molecules to be studied need to be placed in

Maurice Wilkins with a study model of DNA molecular structure. Wilkins's work would earn him a Nobel Prize for Medicine and Physiology in 1962.

an ordered arrangement, more or less lined up with each other in a "crystal" orientation. Randall, Wilkins's boss, knew they'd need an expert in X-ray technology. Randall hired Rosalind Franklin—who had exactly the expertise they needed—and the real soap opera began.

Franklin was one of the world's leading X-ray crystallographers when she began working to uncover the structure of DNA.

NO CLEAR STRUCTURE

No one knows exactly why—although there are a number of possible personal and professional explanations—but Franklin and Wilkins took an immediate dislike to each other. That dislike was the most important random factor that denied Franklin and Wilkins the credit they deserved for the scientific work they did. Wilkins's team was able to crystallize DNA, and Franklin took the measurements.

Wilkins had told Franklin of his suspicion that DNA was helical, and that was enough reason for her to try to prove him wrong. She took X-ray diffraction measurements of DNA, but the extracted DNA had two forms: a "dry" and "moist" form. For Franklin, used to working with inorganic molecules, the "dry" form was the right one to be measuring. Biologists knew that molecules in living organisms were always in a moist environment, but Franklin was not asking the biologists their opinion. She spread the word about the data she'd measured from the dry form, data which showed no clear structure. The data she'd measured from the moist form she kept to herself. The X-ray diffraction picture from the moist form looked like a series of dots in an "X" pattern. To any trained eye, that data shouted out that DNA has a helical structure.

Then, as now, scientists would present their results, ideally sharing successes and failures so each group would learn from another. Watson attended one of Franklin's talks, then excitedly rushed back to Crick to try out new models. Unfortunately, Watson hadn't taken notes, and he misremembered Franklin's comments on the amount of water she'd found in preliminary measurements. Since Watson made some wrong assumptions, the partial model he and Crick developed was obviously incorrect. When Watson took advantage of an opportunity to present the partial model, he paid for his mistakes. Franklin took him to task for his superficial and careless science. Watson's public embarrassment

James Watson and Francis Crick with their DNA model.

at Franklin's hands had consequences. When he returned to Cambridge, Bragg told him enough was enough: he and Crick were not allowed to work any more with DNA. But Bragg didn't give him anything else to do; so Watson kept hanging out with Crick, talking about DNA.

That was the next chance circumstance. With Watson and Crick now officially not working on DNA, Wilkins at King's College felt comfortable talking with his friend, Crick. Wilkins told Crick that his data hinted at a possible helical structure, but that the strand of the molecule was too thick to be a single helix. If he was right, DNA must be a double helical strand, but Wilkins couldn't be sure.

Back at King's College, the soap opera continued. Wilkins ran across a copy of an X-ray diffraction photograph. It was Franklin's data from the moist DNA. Wilkins was stunned. The outline was right there: DNA had to have a helical structure! But the data still belonged to Franklin, and she wouldn't release it. Wilkins was both angered and excited by his find. He talked to someone who happened to be visiting, someone he knew would understand: Watson.

Watson was excited by the news. He ran to the head of the lab, Bragg, and asked for permission to restart their investigation into the structure of DNA. He appealed to Bragg's patriotism. Time was going by. With each month Pauling was

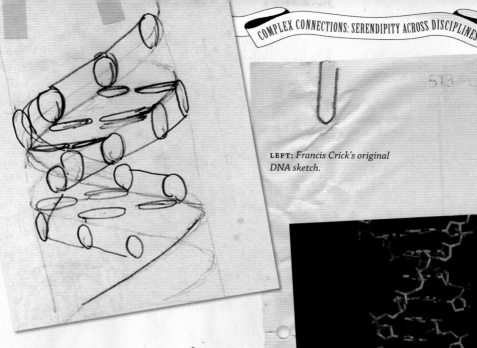

513 Odeor

LEFT: *Francis Crick's original DNA sketch.*

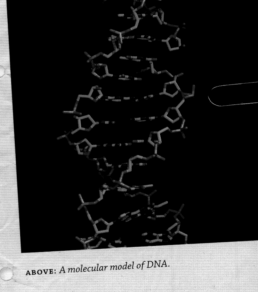

ABOVE: *A molecular model of DNA.*

undoubtedly getting closer to a solution. Surely a little internal competition between British research groups was forgivable if it would increase the chances that Britain would be first with a description of DNA? Bragg agreed.

Watson rushed to Crick. Watson knew of Wilkins's earlier suspicion the molecule was a double helix; now Wilkins had proof. Now that Watson knew for certain that DNA was a double helix, the problem was simpler. How did the sugars and the purines and pyrimidines fit into the double helix structure? The chemistry indicated that the long rails, or backbones, of the helix were made from the sugar molecules connected through short phosphate groups. The two neighboring rails must be connected to each other somehow. There were two possible structures. The two long chains could be attached directly to each other then the purines and pyrimidines would be pointed out like thorns from a stem, or the purines and pyrimidines could connect the two rails together making a twisted ladder. Watson and Crick sat in Crick's office playing with models, trying to get something to fit, but nothing worked.

Watson and Crick despaired. They must be close, but any way they tried to fit the A, T, G, and C bases

along the long sugar "backbones," they created a lumpy, uneven structure that couldn't be physically supported. Watson knew whatever structure he came up with would need to be consistent with the fact that DNA could make perfect copies of itself; so there had to be a simple rule governing the design. The rule would need to relate the A, T, G, or C attached to one sugar backbone to the A, T, G, or C attached to the other at the equivalent location on the other sugar backbone. What could be simpler than requiring each adenine, for example, to match up with another adenine on the opposite chain? Watson convinced himself his still somewhat clunky model was right. Then chance played its final hand.

Jerry Donohue was a chemist visiting from CalTech. In an instance of pure geographic coincidence, Donohue's office was right next to Watson's. Watson proudly described his DNA model when Donohue stopped in. Donohue told him he had it wrong. The purines and pyrimidines are relatively small organic molecules, but they are organic molecules, and one of the complications associated with organic molecules is that they can exist in more than one physical form. All the textbooks said that the purine and pyrimidine bases existed in a configuration called the "enol" form. Donohue had done his own X-ray diffraction measurements of the A, T, G, and C molecules and he knew the textbooks were wrong. Donohue told Watson and Crick those molecules were normally in a different configuration called the "keto" form.

A karyotyped set of male chromosomes, classified and arranged according to the number, size, shape, and other characteristics of the chromosomes.

Watson and Crick made new models of the proper forms for the A, T, G, and C molecules. Watson was disappointed because he just couldn't create a workable model that fit A to A, T to T—the "like-to-like" model just didn't work! He let his model sit overnight, came in the next morning and suddenly the complete DNA model snapped into place. Where one sugar backbone held a guanine purine base it would match up with a cytosine pyrimidine base on the other sugar backbone molecule. The same rule held for adenine and thymine. The two sugar backbones were like the rails of a ladder. Each rung was built out of one specific purine and one specific pyrimidine. Then the ladder was twisted around itself. T and A linked together had pretty much the same shape as G and C linked together. Each rung of the ladder was the same length. T matched to A, G matched to C all the way up the ladder.

Meanwhile, back at King's College, Wilkins was excited. Finally Randall was taking action to eliminate the strife in his lab. He asked Franklin to leave and turn over her data to Wilkins. Wilkins now officially had the X-ray data and he was prepared to take the final steps. He wrote to his friend Crick, telling him that he, Wilkins, would soon have the answer he'd been seeking for so long. Wilkins had no clue that Crick and Watson had the answer, arrived at largely through the information they garnered from the work of Wilkins and Franklin.

Without having spent a minute in the lab, or planning or running a single experiment, Watson and Crick had solved the structure of DNA. To Watson and Crick went the laurels. Although they shared the Physiology or Medicine Nobel Prize with Wilkins, the public acclaim went to Watson and Crick. Some believe Watson and Crick deserve their recognition because of the insight they brought, some feel the recognition is misplaced because others planned and carried out the experimental plan that was certain to reveal the structure sooner or later. One thing is certain. Without unique accidents of personality and geography, Watson and Crick would not today be celebrated as the pair who cracked the secret structure of DNA.

Truly insightful?

The argument still rages today over how much true insight Francis Crick and James Watson demonstrated when they deduced the structure of DNA. But determining the structure of DNA was only a preliminary step toward answering the real question: how did such a chemically simple molecule perform such a biologically complex task? The molecule responsible for carrying genetic information needed to have some mechanism for creating perfect copies. One prominent proposed mechanism held that there was some template molecule that would sit in a central location making copies of itself and sending them out into the cell.

Watson and Crick thought about the structure they had discerned and recognized that the DNA molecule was both the template for copying the genetic information and the genetic information itself. Just one month after they published a paper describing the DNA structure, Watson and Crick published a paper describing how DNA could be duplicated. Because each A would only bind to a T and each G to a C, the two DNA strands were complementary. That is, if the strands were unwound and put in an environment where A, T, G, and C bases were floating around, each unwound strand could automatically make a complementary copy of itself. It is an elegant mechanism responsible for life on Earth. The order of the A, T, G, and C bases is preserved. Later it would be learned that a complicated set of biological machines work to translate the order of the bases into proteins that build and run the bodies of living organisms on Earth. Watson and Crick made the most out of their accidental opportunity, and each did work that extended our knowledge of human biology.

KEEPING THE BEAT

★

IN 1958 CARDIAC SURGEONS *had a problem: their open heart surgeries were successful. Blockages were fixed, arteries were bypassed, and yet. . . patients were dying. Nerves that send signals to direct the heartbeat are not readily visible. Surgeons had a general idea where the nerves were, but slight differences in nerve location from patient to patient meant surgery to move an artery could end up damaging the nerves. The damage could often go undetected until hours or days after surgery, when a patient's heart would fail. Nearly ten percent of surgery patients would die from this nerve damage, called "heart block." Diseases and congenital defects could also lead to heart block, then failing heartbeat, and death. An engineer's mistake was about to fix the problem.*

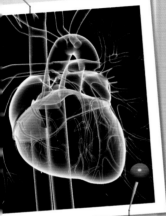

A lifeline in an electrocardiogram.

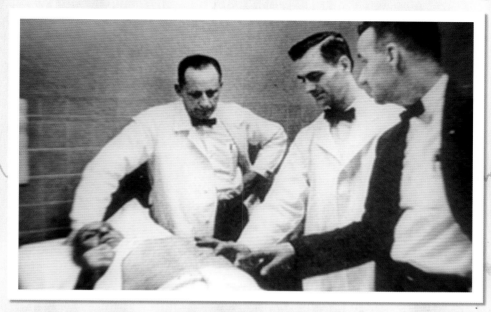

The "Bow Tie Team" which developed the implantable type of pacemaker. From left: Dr. William Chardack, Dr. Andrew Gage and Wilson Greatbatch.

LUNCHING WITH SURGEONS

After serving in the US Army Air Force in World War II, Wilson Greatbatch went to Cornell University. He supported his education and his growing family by constructing and maintaining equipment at the local radio station. In 1951 he was working at the Cornell Psychology Department's animal behavior farm. He built and maintained instruments to monitor physiological parameters such as brain waves, blood pressure, and heart rate. For a time he spent his lunch hour with surgeons visiting from New England. The surgeons described heart block, and Greatbatch realized it sounded like a problem he was familiar with. It was like a failure in radio communications. The signal was being sent, it

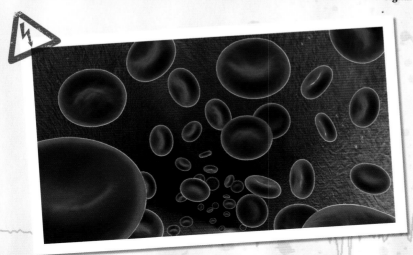

Red blood cells moving through the blood stream.

just wasn't being received. He'd solved problems like that before; so he knew he could solve this problem, if he had the tools.

But he didn't have the tools.

IN A HEARTBEAT

Greatbatch's career took him to Buffalo, where, in 1958, he was teaching electrical engineering at the University of Buffalo. He also worked for the Chronic Disease Research Institute. In April, 1958, Greatbatch was using the newly available silicon transistors to build an oscillator circuit for measuring fast heartbeats. He reached for a 10,000-ohm resistor, a small cylindrical component marked with a brown, black, and orange band. Instead, he grabbed a resistor marked with a brown, black, and green band—a 1,000,000-ohm resistor. He completed the circuit and tested it. Something was wrong; he had made a mistake. Instead of producing a single continuous tone, his circuit was "squegging," putting out a very short pulse, shutting down for about a second, then putting out another short pulse.

Greatbatch stared at the circuit. He immediately thought back to his discussion of seven years ago. Heart block happened when an electrical signal didn't make it to the muscles of the heart. Greatbatch had developed a compact transistorized circuit that generated exactly the kind of signal the heart was missing. Three weeks later Greatbatch brought his device to Buffalo's Veterans Administration Hospital, where surgeon William C. Chardack tested it by implanting it in the body of a dog suffering from heart block. Chardack connected wires to the dog's still heart, and it began to beat. The signals to control the heartbeat were not coming from the brain through the nerves, but directly from Greatbatch's electronic pacemaker. Chardack summarized their thoughts: "I'll be damned," he said.

A SIMPLE MISTAKE

The first implantable pacemaker was in use. Greatbatch knew his electronics, but before this he'd never made anything to be implanted in a body. The device, wrapped in electrical tape, failed in four hours. But it had worked. Greatbatch knew he was on to something; so he quit his job and devoted himself to improving the pacemaker.

His next pacemakers were built in a solid block of epoxy, and they lasted up to four months. Over the next two years Greatbatch refined his design, and on April 15, 1960, Chardack's group implanted 10 pacemakers in patients with complete heart blocks. Without assistance, they had about a 50 percent chance of surviving for a year. The first patient lived for 18 months. The second lived for 30 years.

Greatbatch spent years improving and refining the design. Soon pacemakers were reliable enough to last for several years. They needed to be replaced not because their electronics failed, but because the batteries ran out. Greatbatch spent years on that problem, and developed a lithium battery that can work for up to ten years. Due in large part to Greatbatch's simple mistake, he devoted a career to developing and refining a device that at this moment is keeping more than 10 million people alive.

ABOVE: *Cardiac pacemakers can now run on lithium batteries for up to ten years without failing.*

HOW DO YOU HEAL A BROKEN HEART?

At the same time as Greatbatch was developing his device, two independent groups were working on the same problem. One prominent team was the combination of engineer Earl Bakken and surgeon Walton Lillehei in Minnesota. To stabilize patients after open heart surgery, in 1957 Lillehei had developed a procedure for leaving wires attached to the heart and hooking patients up to a cabinet-sized electrical generator until their hearts recovered. But a power failure after surgery caused the death of a patient, and Lillehei asked Bakken to develop a portable solution. Bakken modified an electronic metronome circuit small enough to be "worn" by a patient, and brought it in to the hospital where Lillehei verified its effectiveness on a dog. Visiting the hospital the next day, Bakken was surprised to see Lillehei was already using it on a post-operative patient. When Bakken questioned it, Lillehei said he didn't want to lose another patient by failing to use the latest technology.

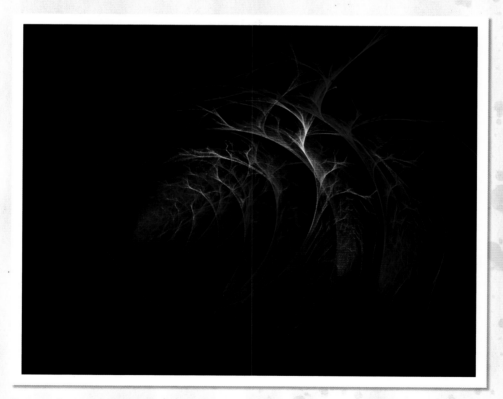

Heart attacks occur when the supply of blood and oxygen to an area of heart muscle is blocked, usually by an arterial clot.

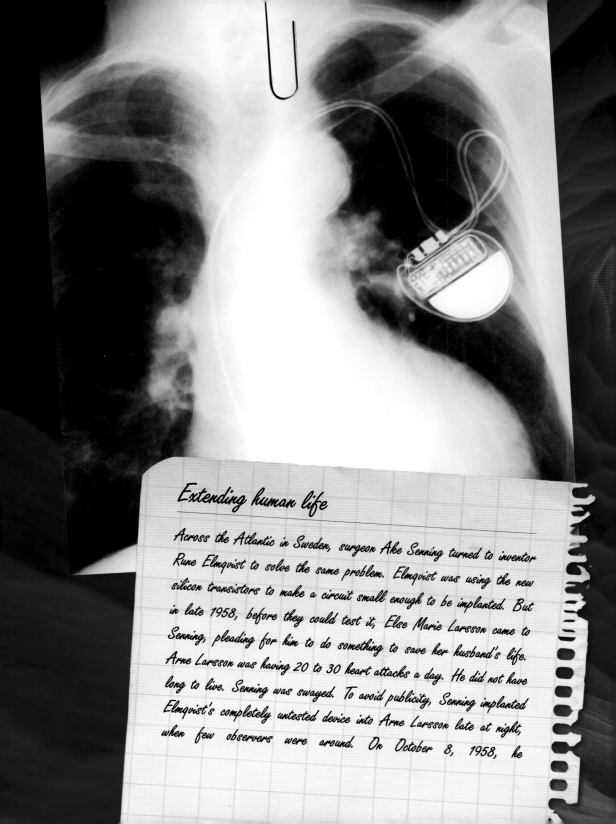

Extending human life

Across the Atlantic in Sweden, surgeon Ake Senning turned to inventor Rune Elmqvist to solve the same problem. Elmqvist was using the new silicon transistors to make a circuit small enough to be implanted. But in late 1958, before they could test it, Else Marie Larsson came to Senning, pleading for him to do something to save her husband's life. Arne Larsson was having 20 to 30 heart attacks a day. He did not have long to live. Senning was swayed. To avoid publicity, Senning implanted Elmqvist's completely untested device into Arne Larsson late at night, when few observers were around. On October 8, 1958, he was the first human to receive an implanted pacemaker. It failed eight hours later. But Elmqvist had another ready. That one lasted a week. In the remainder of his 86 years, Larsson had 22 different pacemakers implanted. He died in 2001 from causes completely unrelated to his heart disease.

Bakken, Elmqvist, and Greatbatch started the era of electronic augmentation of human organ function, creating the expectation that technology can be harnessed to extend human life.

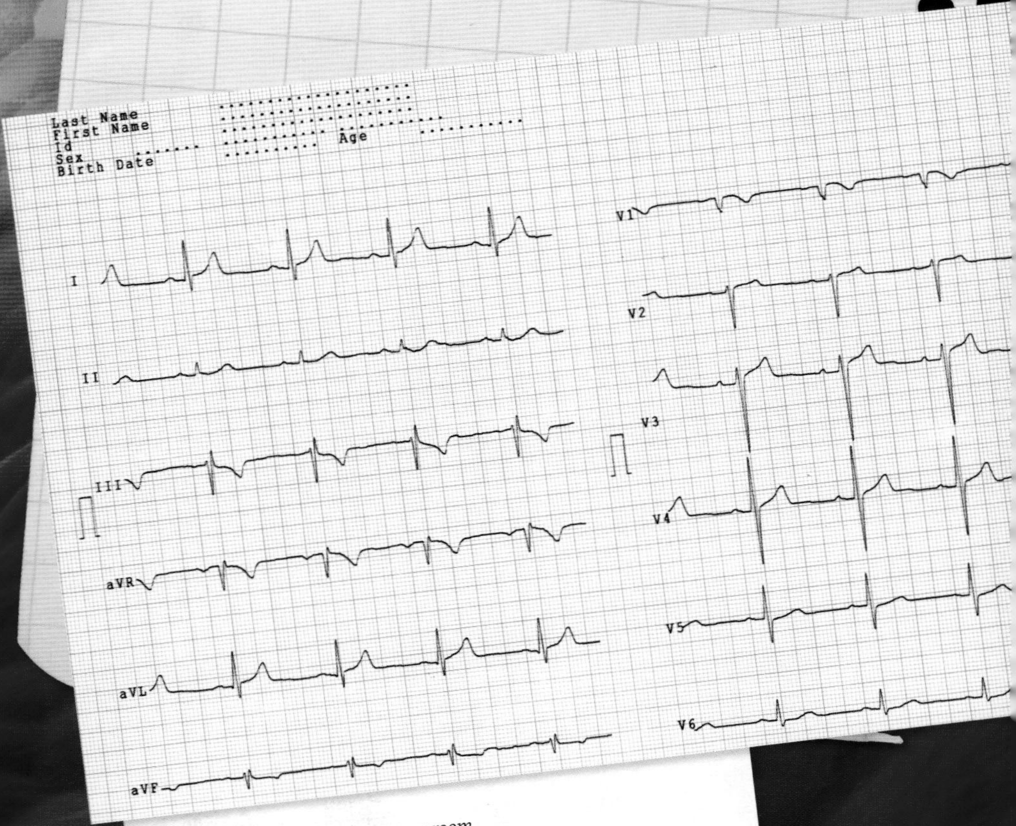

Last Name
First Name
Id
Sex
Birth Date
Age
I
II
III
aVR
aVL
aVF
V1
V2
V3
V4
V5
V6

An ECG taken in an emergency room.

STEPHANIE KWOLEK

INVENTS DuPont™ Kevlar®

1965

DON'T GUM UP THE WORKS!

★

THE AGE OF SYNTHETIC *materials was well advanced in the 1960s. People were no longer surprised to learn of a new durable fabric or flexible plastic. Chemical companies had entire departments whose only function was to invent new materials. Yet even in a place whose purpose was defined by the search for the unexpected, there was still room for surprise.*

SEM (scanning electron microscope) image of the fabric Gore-Tex.

NEVER LOOK BACK

Stephanie Kwolek got her degree in chemistry to prepare herself to be a doctor. But she needed a job to pay for medical school. In 1946, she interviewed for a job at DuPont. Her interview went well, but when told they'd get back to her, she said she needed an offer right then. They offered; she started working for DuPont, and she never looked back.

In 1964 Kwolek was working in the Pioneering Research Laboratory of the textile department at DuPont. This department had made chemical discovery into, well, into a science. Chemists would select likely candidate molecules—monomers—and combine them to produce a polymer. They would then melt the polymer and send the melted polymer to be spun. The spinning department took the melted polymer and funneled it through a spinneret. The spinneret lined up the long polymer molecules, creating a single long fiber. The fiber would then be sent to a testing department, where the physical properties would be measured. The entire system was set up to streamline the development of synthetic fibers. But even in such a streamlined environment, things didn't always run smoothly.

AS STRONG AS STEEL

In the early 1960s the Pioneering Research Laboratory was searching for a specific product: a lightweight, high-performance fiber that could replace steel belting in automobile tires. The DuPont scientists realized that sooner or later automobile gas mileage would become important. If a material could be found that was as strong as steel, but lighter, then mileage could be improved without sacrificing performance.

Kwolek had helped to create Nomex, a polymer resistant to high temperatures. Today Nomex is an important component of protective equipment for firefighters and is also commonly used for electrical insulation. Kwolek believed a monomer similar to that which composed the Nomex polymer could also be used to build a high-strength fiber. But she ran into a problem. The polymer she made wouldn't melt. A polymer that wouldn't melt couldn't be pulled into a fiber. Since it wouldn't melt, Kwolek tried another approach. She mixed the polymer with a solvent and it liquefied.

But there was a problem. Liquid polymers were supposed to look something like maple syrup, thick

Firefighters' uniforms need to be made from materials that are heat resistant and flame retardant.

RESISTANT

and translucent. This one looked like cloudy water. Kwolek took her liquid polymer to the spinneret, but the folks in charge didn't want to put it through their machine. They thought the liquid Kwolek had brought them was contaminated with particles that would clog the machine. Kwolek filtered her material and determined that it wasn't made from solid particles suspended in a liquid, but a single type of liquid polymer. She brought it back to the spinneret. Still they didn't want to risk damaging the equipment. She wouldn't give up, though, and eventually they put the cloudy liquid through the spinneret, where it spun beautifully. Finally, Kwolek had her fiber. She brought it to the testing department.

She didn't believe the results.

PREPARATION AND DESIRE

According to the tests, the fiber was nine times stiffer than anything she'd made before. It was pound-for-pound five times stronger than steel. These were amazing results, as good as anything Kwolek had hoped for, but were they right? She sent the fiber to be tested again. And again. Finally she was convinced. When she told her colleagues there was immediate excitement. They realized the potential of the discovery.

Kwolek had faced down questions about the compound she had developed by running tests to convince herself it was worth looking into. She later determined that the cloudy appearance of the liquid was because her polymer was turning itself into a liquid crystal. Most polymer molecules, although they're long strings, will twist and wind a little, and maybe even put out little branches in different directions. The polymer molecules she had made were almost perfectly straight. When she dissolved them into a liquid, in some places the molecules lined up with each other and in others they crossed each other, giving her liquid that cloudy look.

It's a good thing Kwolek persisted, because the strong fiber she found is now known as Kevlar®. Kevlar® appears in space suits, suspension bridges, automobile and airplane brake pads, skis, boats, safety helmets— any place where strength is needed in a lightweight package. Kwolek had found the unexpected while looking for the unknown, and her preparation and desire pushed her to take advantage of her opportunity.

Various products and materials are produced using DuPont™ Kevlar® technology.

Safe and secure

The most famous application of DuPont ™ Kevlar ® and other fibers of the same chemical class is in bulletproof vests. A bulletproof vest consists of several layers of Kevlar ® fibers, with each layer consisting of interwoven or laminated fibers. When a bullet strikes the vest, the fibers in a single layer are strong enough to spread the impact out before they break. If the bullet is powerful enough to break through the first layer, the second layer acts the same way. By the time the bullet reaches the last layer——usually the seventh——the force is spread out so much that the bullet doesn't have enough energy left to penetrate the body.

When designers at the US Army's Edgewood Arsenal were developing the first Kevlar ® bulletproof vest, they ran it through a series of tests. They determined that the penetration resistance of the fiber was diminished if it got wet, was exposed to ultraviolet light, if it was bleached, or even after repeated washings. Because of that, bulletproof vests protect the Kevlar ® against exposure to light and water. Kwolek's cloudy solution started as an ugly duckling in the laboratory, but to the thousands of police officers whose lives have been saved due to Kevlar ® vests, it is a beautiful, life-saving swan indeed.

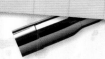

S. JOCELYN BELL

DISCOVERS PULSARS

1967

RECEIVING THE LONG-DISTANCE CALL

MORE THAN IN ANY *other field of scientific inquiry, chance is critical in astronomy. There are no laboratory benches, no experiments: astronomers look at the sky and see whatever the universe offers up. But even in astronomy, some discoveries are more accidental than others. Jocelyn Bell (later Jocelyn Bell-Burnell) was applying the tools of radio astronomy——the field that sprang from Karl Jansky's accidental observation——to a specific problem. Instead she found the unexpected and started a chain of inquiry that expanded our understanding of the nature of matter.*

Communication antennas receive and transmit radio signals from satellites orbiting the Earth and pass this information on.

Radio signals get broken up when they travel through the cloud of charged particles surrounding the sun.

HOW BIG ARE QUASARS?

Antony Hewish developed expertise and interest in radio technology during World War II. After the war he became a radio astronomer at Cambridge University. In 1967, Hewish wanted to answer a simple question: how big are the bright, distant radio sources known as quasars? He planned to do this by observing scintillations, rapid changes in the strength of the radio signal.

Stars "twinkle," or scintillate, because they are so small that their light gets kind of broken up going through the atmosphere. In the same way, a small radio source gets broken up when the signals travel through the cloud of charged particles surrounding the sun. To measure this radio "twinkle," Hewish designed the Interplanetary Scintillation (IPS) Array, an antenna consisting

of about 124 miles (200 km) of wire hanging on 1,000 posts. To catch twinkling sources, the IPS Array was designed to capture intensity changes as short as one-tenth of a second. Just that characteristic would lead to an accidental discovery.

AS THE EARTH ROTATED

In 1967 Jocelyn Bell was a PhD student studying under Hewish. She had helped build the IPS Array, and as it neared completion she became the person in charge of operating it. The antenna could be electronically "pointed" anywhere along a north-south line. As the Earth rotated, a source would only stay in view for

Dr. Anthony Hewish, radio physicist, with young research assistant, Jocelyn Bell.

location as the earlier scruff, in the constellation Vulpecula. She made more measurements in that location, and again the scruff appeared.

Bell's scruffy signal was unusual for more than the way it looked. First of all, as the Earth rotated, the telescope measured a particular slice of sky for about four minutes, but the scruff only lasted for a minute. Another strange thing: the signal first appeared in the middle of the night, with the telescope pointed far away from the charged particle cloud surrounding the sun. By October, Bell had gathered enough data to be convinced she had found something interesting.

LITTLE GREEN MEN

Bell took the data to Hewish, who had her do a faster measurement of the source. "Faster measurement" meant turning up the speed on the chart paper, which then had to be turned down afterward to avoid filling the lab with paper. Of course that task fell to Bell. Not only was it an onerous task, but it wasn't all that rewarding, because the signal disappeared for weeks. Finally, on November 28, 1967, Bell recorded the most unusual radio signal yet detected. The source put out a pulse of radio energy less than a tenth of a second long, went quiet, then sent out another pulse 1.3 seconds after the first pulse. It repeated that for about a minute. Sometimes the pulse was intense, sometimes dimmer, but always at a precise interval (which has now been accurately measured at 1.3372795 seconds). Bell grabbed Hewish for the next day's measurement and, luckily, her pulsing radio source cooperated.

Hewish and Bell speculated about what could be causing the signal. It was precisely regular. Was it a signal from some advanced civilization? Bell half-playfully labeled the source "LGM1," for "little green men one." Bell went back to her charts and found even more scruff, which she confirmed as three more sources of pulsing radio signals in different parts of the sky

about four minutes. Bell's job was to direct the antenna to scan the sky, recording the antenna signals on rolls of chart paper. The paper unrolled about 98 ft (30 m) a day as pens moved up and down depending upon the signal strength. It was Jocelyn Bell's job to look at all that paper. She got pretty good at her job.

Bell had started taking data in July of 1967, before the antenna was even completed. She was looking for rapid, irregular changes in signal intensity—the kind of signal a scintillating point source would make. She became familiar with the look of background interference from mundane sources: transmitters, motors, automobiles starting—just about anything electrical can send out radio signals. About six weeks into her measurements she noticed something that was neither scintillation nor Earthly background. She called it "scruff."

At first, she just noticed the scruff, made a notation, and went about her task. But some time later she noticed more. It was coming from the same

Jocelyn Bell and Dr. Anthony Hewish monitored printouts of information received via the antenna.

NEUTRON STARS

Eventually the explanation arrived. "Rapidly pulsating radio sources"— or pulsars, as they were soon called— are neutron stars. Neutron stars are the remnants of supernovae, matter so dense it's as if our sun was compressed down to the size of a mountain. Some neutron stars are rotating and they have magnetic fields strong enough to whip charged particles around themselves as they spin. Some of that energy gets released in the form of radio waves, which sweep across the sky like a beam from a celestial lighthouse. Neutron stars were postulated before the detection of pulsars, but without that experimental confirmation they would have remained just a theoretical construct. The confirmation of the existence of neutron stars was also the confirmation of our understanding of nuclear physics, explaining the structure of everything we'll ever touch. In addition, pulsars themselves are a tool for probing the mechanisms of gravity itself, the force that holds the universe together.

Jocelyn Bell was open to the chance observation of "scruff," leading to a range of studies that have expanded our understanding of our world on both the smallest and the largest scales.

with different pulse frequencies. Unless the interstellar phone rates were at a low six-month introductory rate it was pretty unlikely that several different civilizations were trying to talk to the universe at the same time.

Pulsing radio signals in different parts of the sky led to the confirmation of the existence of neutron stars.

Listening to the radio

One of the most famous examples of an accidental radio antenna observation happened in 1964. Bell Laboratories had a radio antenna they used for satellite communications at their site in Holmdel, New Jersey (the same location as Jansky's measurements, but a different antenna). The antenna was extremely sensitive; so when Bell Labs decided not to use the antenna to communicate with satellites, Arno Penzias and Robert Wilson wanted to refit it for use in radio astronomy. Before they could use it, they needed to identify and eliminate all the sources of error. They meticulously accounted for and eliminated every source of noise, up to and including pigeon poop on the antenna.

But they were still left with a low level of radio background that filled the sky. Frustrated, Penzias talked with a colleague, MIT researcher Bernard Burke. Burke happened to be familiar with work being done by a group at Princeton. The Princeton group had predicted that if the Big Bang theory of the formation of the universe was correct, there should be a leftover low level of radio energy. Penzias got in touch with the Princeton group. The Princeton folks came out to take a look at Penzias and Wilson's data, and the noise was explained. Penzias and Wilson were measuring the remnants of the explosive beginnings of the universe. It was an entirely accidental observation and it won them a Nobel Prize.

ALAN MACDIARMID,
HIDEKI SHIRAKAWA,
AND ALAN HEEGER

INVENT AN INTRINSICALLY
CONDUCTING POLYMER

1977

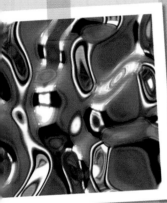

PLASTICS GET CURRENT

IN THE 1970S *plastics were recognized for being a lot of things: flexible, lightweight, moldable, inexpensive——but one thing they weren't recognized for was being electrically conductive. Alan MacDiarmid was trying to change that. He wanted a material that was as versatile as plastic, but conductive like a metal. He had a plan for getting there, but without a laboratory accident and some serendipitous timing, he might never have arrived.*

PHYSICS

CHEMISTRY

TECHNOLOGY

A GOLDEN LUSTER

In 1975, Alan Heeger was studying interesting questions about the metal-insulator transition. Every atom is composed of a cloud of electrons surrounding a small core of larger protons and neutrons. In metals, when the atoms are put together some of the electrons from each atom are "cut loose," they just kind of float around everywhere in the material. Because the electrons are not held in a specific location, they flow very easily, creating an electrical current. In insulating materials, the electrons remain stuck around the individual atoms; so they don't flow. Some metals have very high conductivity; that is, they're very metallic. Some insulators have very high resistance; that is, they're very insulating. But some are kind of in the middle. The metal-insulator transition refers to the atomic or molecular structure that is right on the cusp between conducting and non-conducting, where a tiny, subtle difference in a molecule puts it over the edge in one direction or another. Heeger, a physicist, needed some help from a chemist; so he sought out his University of Pennsylvania colleague, Alan MacDiarmid.

MacDiarmid began working on questions of conductivity, measuring different materials, evaluating their structure, trying to determine what made one particular material a conductor while another would be an insulator. He had developed a poly-sulfur-nitride polymer that exhibited some conductive properties. It also had a golden luster. MacDiarmid presented his work at a seminar at the Tokyo Institute of Technology. His talk was about conductivity in silicon-like materials and plastic-like polymer materials. Hideki Shirakawa was a polymer chemist at the Tokyo Institute; so he would have been interested in MacDiarmid's talk, but the notice had printed the title of MacDiarmid's talk with "silicon" part in large type and the "plastic" part in small type. Shirakawa missed the title, and missed the talk. Luckily, the seminar organizer recognized that Shirakawa would be interested; so he tracked him down after the talk. At the coffee break (actually, a "green tea break") Shirakawa introduced himself to MacDiarmid. And Shirakawa didn't come empty-handed.

MIXING WITH A CATALYST

Shirakawa's lab produced a variety of different polymers in an attempt to understand how their chemical structure influenced the properties of the bulk material. One of the compounds Shirakawa was working with was polyacetylene. A visiting Korean scientist had been working in Shirakawa's lab, and Shirakawa had directed him to make polyacetylene by mixing chemicals with a catalyst.

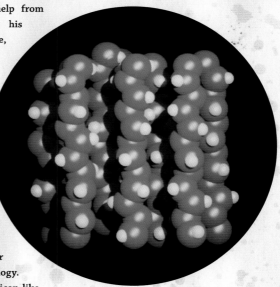

The high electrical conductivity discovered in polyacetylene polymers accelerated interest in the use of organic compounds in microelectronics.

Doped polyacetylene is a little like a chain of dominoes: the bonds can flip all the way up the polymer molecule.

A SILVERY LUSTER

But the language barrier created a problem. In one of the happier instances of miscommunication, the Korean assistant put in 1,000 times more catalyst than Shirakawa had directed. Instead of a black powder, he created a thick, silvery, lustrous goop. Shirakawa had set the beaker aside as an item of interest, an item that he then realized would interest MacDiarmid as well. MacDiarmid took one look at Shirakawa's polyacetylene; saw the silvery luster—similar to the golden luster of the poly-sulfur-nitride he had made—and knew there was something worth following up. But if he were to investigate properly, he would need someone with polymer expertise. On the spot, MacDiarmid invited Shirakawa to join him at the University of Pennsylvania.

A CONJUGATED POLYMER

Heeger, MacDiarmid, and Shirakawa worked with polyacetylene, adding impurities in the same way that impurities are "doped" into semiconductors to change their properties. Polyacetylene is what's called a conjugated polymer, which means there are alternating single and double chemical bonds throughout the length of the material. By doping polyacetylene, MacDiarmid's group created a situation where the single and double chemical bonds could exchange places. Like a chain of dominoes, the bonds can flip all the way up the polymer molecule. In chemical terms, the single and double bonds change places, but in physical terms a current is moving along the polymer. The researchers had turned a polymer plastic into an electrical conductor, and changed our understanding of materials in a fundamental way.

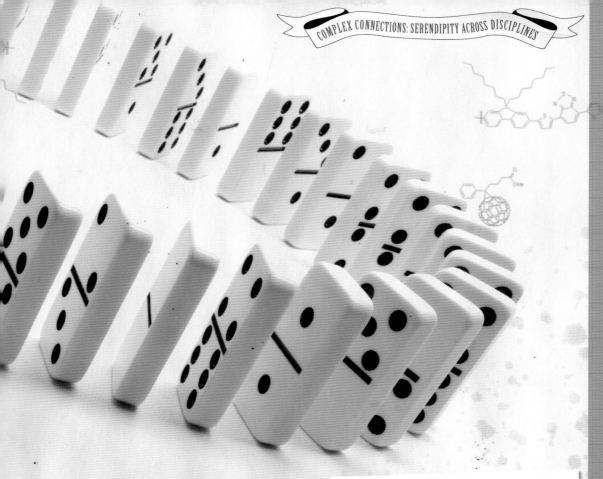

MacDiarmid, Shirakawa, and Heeger had opened up an entirely new field, creating a new class of materials with the electrical and optical properties of metals but the processing advantages and mechanical properties of plastics. The three researchers received the Nobel Prize in Chemistry in 2000. The Nobel Committee was impressed with both the scientific importance of conductive polymers as well as their potential for practical applications. However these essential materials might never have been developed without an accidental misunderstanding, the preparation to recognize something unusual, and the desire to probe the unknown.

Hideki Shirakawa made a press declaration after winning the Nobel Prize in Chemistry in October 2000 along with Alan Heeger and Alan MacDiarmid.

Polyacetylene polymers are used in microelectronics because they are electrically conductive.

Coming from every angle

Shirakawa, Heeger, and MacDiarmid were facing a complex problem with technical and scientific issues in chemistry, physics, and materials science. In previous decades, researchers would have addressed such questions by filling in the gaps in their own knowledge. When Shirakawa, Heeger, and MacDiarmid did their work, that was no longer practical. Too much groundwork needed to be done to establish even a basic understanding of different fields. So, when MacDiarmid ran into Shirakawa and recognized he needed polymer chemistry expertise to succeed, it was logical to seek that expertise by simply bringing in a collaborator.

The three learned lessons about what is required for a successful scientific collaboration. Each member of the team needed enough in-depth preparation to be able to identify unusual results related to their expertise. But they also needed to learn the specialized language of each discipline—for example, chemists and physicists use the same terms, but they mean something different. Then the three researchers needed to broaden their understanding enough to provide insights outside their own specialties. The process itself was challenging and remarkable enough that the Nobel Award Citation specifically mentions advances in "interdisciplinary development between chemistry and physics."

CHAPTER 5

THE CORPORATE MODEL: CHANCE IN BIG SCIENCE

★

A CINDERELLA STORY
Harry Kroto, Richard Smalley, and Robert Curl discover fullerenes, 1985

RUNNING COLD, RUNNING FREE
Alex Müller and Georg Bednorz discover High-Temperature Superconductivity, 1986

SPINNING A WEB
Tim Berners-Lee creates the World Wide Web, 1990

JUST SPINNING THE WHEELS
The Spirit Rover uncovers evidence of free-flowing water on Mars, 2005

The late twentieth century saw an evolution in the model of scientific research. As ever, the pace of specialization was increasing, but there was a parallel realization that interdisciplinary research could lead to big scientific and technological payoffs. Scientific research—whether in industry, the public sector, or academia—increasingly began to follow the corporate model. In the corporate model, research is targeted. There is a goal. Goal-oriented research is almost the antithesis of discovery-driven science. Within this framework, chance still plays a big role, but recognizing and acting upon chance observations requires scientists to demonstrate personal commitment and courage.

This is obvious in a corporate setting. If, for example, a corporation wants to develop a heat-resistant coating, the administration will evaluate proposals, allocate funding, and track progress toward that goal. A researcher working that task could, say, stumble across an electrically conductive transparent thin film, but to pursue that lead requires the researcher to go back up the chain and explain why the milestones on the program are not being met. Even if the scientist is successful at shifting the corporate priority, the new research now needs resources to be allocated, and it gets its own new budget and schedule. Not necessarily conducive to open scientific inquiry, of course. But the corporation may not care about open scientific inquiry; they just want their heat-resistant coating! But it would be difficult to argue that significant discoveries are not being overlooked in corporate research environments.

The situation is different in academia, but perhaps not as different as might be expected. Today's university research is funded primarily through grants. And almost all grants have goals attached. Whether public or private, funding agencies want to see results. Grants are given for specific projects. It would be difficult for a researcher working under a grant entitled "Role of Mineralocorticoid Receptor in Diabetic Cardiovascular Disease" to follow a chance observation and discover a new method for growing retinal cells. After one project is complete, researchers need to move onto the next. If they haven't demonstrated progress—if they have been sidetracked and distracted by a chance observation—it's going to be a bit tougher to obtain follow-on funding. Chance still has a role, it's just that scientists always need a little bit of courage to follow up an accidental observation.

It takes more than courage, though, because most science is now done in teams. It takes persuasive ability as well. Two of the accidental discoveries described in this chapter would not have come to fruition without the active approval or involvement of large scientific and technical teams. The remaining two didn't require as many people to be convinced, but they still needed active involvement of scientists on the research teams. If the fire of discovery can't be lit in the people who must be convinced, discoveries languish.

Preparation is still important—and pretty well addressed by the educational structures in place. Opportunities still arise with chance observations. Desire, though, must come from the individuals. If the desire isn't strong enough, opportunity will not lead to discovery. The examples in this chapter demonstrate that the quest to understand creates strong desire: chance discovery remains alive.

HARRY KROTO, RICHARD
SMALLEY, AND ROBERT
CURL

DISCOVER FULLERENES

1985

Molecules in interstellar
dust absorb starlight on its
way through.

A CINDERELLA STORY

IN 1985 HARRY KROTO at the University of Sussex had made a rather interesting discovery. Kroto was doing molecular spectroscopy—predicting and measuring the interaction of electromagnetic energy with molecules. Light from distant stars travels through thin clouds of interstellar dust. The distances are so great that light can be completely absorbed by the molecules in the dust. Kroto had found evidence of the presence of particular long-chain carbon and nitrogen molecules called cyanopolyynes in the interstellar dust. But other scientists were skeptical of his results; so he recruited help to perform an exotic experiment: recreate conditions within a dying star to prove the cyanopolyynes could be produced. But Kroto and his colleagues accidentally produced another molecule, one with a structure that had not been seen before. A molecule with the potential to change nearly every facet of human technology.

A SILLY IDEA

Before 1985 Richard Smalley and Robert Curl had developed a technique for creating and studying "clusters," bunches of atoms latched on to each other. Smalley's technique was to slam a short pulse of high-intensity laser light onto a suitable target. The laser ablates the surface, creating a tiny cloud of loose atoms which are free to interact with each other and clump together. The particles accelerate into a detector. Lighter molecules travel faster; so by timing the arrival at the detector, Smalley could measure the mass of the clusters formed in the cloud of ablated molecules.

Kroto heard of the technique and asked to collaborate on an experiment to recreate the stellar atmosphere of a red giant. At first Smalley thought it was a silly idea. He was more interested in the clustering of silicon and germanium, elements important to semiconductor production. It took a year and a half for Kroto to convince Smalley to let his valuable instrument be dedicated to an experiment with such a small payoff.

Laser beams are pulsed onto a target, ablating the surface and creating a cloud of loose atoms which become free to interact with each other—clusters can then be studied.

LIKE SOOT FROM A FLAME

Finally, in 1985, Smalley and Curl set up a graphite target for Kroto's experiment. Graphite is made from flat sheets of carbon atoms connected in a hexagonal pattern, with the sheets loosely

Graphite is made from flat sheets of carbon atoms, connected together in a hexagonal pattern.

connected to each other (which is why graphite is used for pencils, because the layers will just glide off when rubbed against something like a sheet of paper). The laser created a puff of carbon atoms like soot from a flame; then a stream of helium gas cooled the sooty puff, allowing the carbon atoms to connect with each other in various clusters. In the initial experiment they measured clusters with masses in the range of 38 to 120 carbon atoms. There was a slight peak at a mass of 720 atomic mass units (amu), which corresponded to 60 carbon atoms, and a slightly smaller peak corresponding to 70 carbon atoms. The researchers figured that the laser had broken off chunks of carbon rings from the graphite, which then combined with each other to make the clusters.

The next step was to slow down the measurement, see if any of the clusters created stable molecules. The researchers did that and most of the clusters they had measured no longer appeared. But the peak at C_{60} (the chemical notation for a molecule with 60 carbon atoms and nothing else) was huge, with a lesser peak at C_{70}. This was a puzzle. Carbon is friendly; it reacts easily with other atoms. If there was a 60-atom chunk of graphite, the carbons at the edges would grab onto others and the chunk would grow. The same for a 60-atom carbon chain: the atoms on the end would grab others and keep growing. The only way a 60-atom cluster could be stable would be if every atom was connected to other carbons, but there was no known structure of carbon that could connect 60 atoms to each other with no loose ends.

A BOLD THEORETICAL LEAP

Kroto and Smalley then made a bold theoretical leap. What if their creation was unlike any other molecule ever created: what if it was a sphere? Smalley and Kroto recall the next step in different ways—working in a close team it can be hard to remember which team member did what—but they made several associations. They recalled the image of R. Buckminster Fuller's geodesic domes, such as the one at *Expo '67* in Montreal. They remembered a mathematical proof from the Swiss mathematician Leonhard Euler, showing that a three-dimensional construction made from flat pieces could produce a fully closed surface if it was built out of exactly 12 pentagons filled out with hexagons. The final clue to the mental puzzle: a soccer ball. A traditional soccer ball is built out of 12 pentagons and 20 hexagons. Smalley and Kroto realized that if every meeting point of a soccer ball's pentagons and hexagons were replaced by a carbon atom, there would be exactly 60 atoms, forming a closed surface with no dangling chemical bonds. They named the new molecule buckminsterfullerene, or "buckyball" for short. They also proposed that the larger molecule they had detected, C_{70}, was a hollow spheroidal molecule, which they christened a "fullerene."

ABOVE AND RIGHT: *If every meeting point of a soccer ball's pentagons and hexagons were to be replaced by a carbon atom, there would be exactly 60 atoms forming a closed surface with no dangling chemical bonds.*

ABOVE: *Buckminster Fuller stands in front of his creation, a geodesic dome which acts as the US pavilion at the 1967 World's Fair.*

Professor Richard Smalley, co-discoverer of fullerenes.

The structure they proposed was a radical departure from anything known in nature. There were already two completely different forms of carbon-only compounds: diamond and graphite. But both diamond and graphite have no fixed size, there's always room for one more carbon atom. Not so in a buckyball. The structure Smalley and Kroto proposed was simple and elegant, but was it right? It was five years before scientists were able to make enough of the C_{60} molecule to be able to confirm the structure was exactly the buckminsterfullerene Smalley and Kroto had hypothesized. Since then, scientists have made fullerenes with up to hundreds of carbon atoms.

It's very difficult to convey the revolutionary implications of their discovery. The situation is similar to the birth of organic chemistry, where a new understanding of how carbon atoms form molecules led to the development of fabrics, pharmaceuticals, plastics, and also paints. Every facet of human enterprise from playing at the beach to exploring the surface of another planet has been affected by our understanding and application of organic chemistry. Fullerenes are just as dramatic a discovery. They have so many unique characteristics that it would be a surprise if they didn't revolutionize computing, telecommunications, medicine—just about every human endeavor.

WHERE NO ONE HAD GONE BEFORE

Although fullerenes weren't discovered until 1985, people had been making them since—well, since the days of cavemen. When carbon is released from burning organic material, then wafts through the flame, the conditions are just right to create the hollow carbon molecules. But no one had sifted through the ashes to find them. Smalley, Kroto, and Curl weren't the first to have the opportunity, but they were the first to pursue it. Their preparation let them understand they had made something unusual, and their desire to understand the unknown pushed them to make the discovery.

Fullerenes are a previously unknown configuration of matter with the potential to change every aspect of human technology.

The discovery of fullerenes has revolutionized the development of fabrics, pharmaceuticals, plastics, and paints.

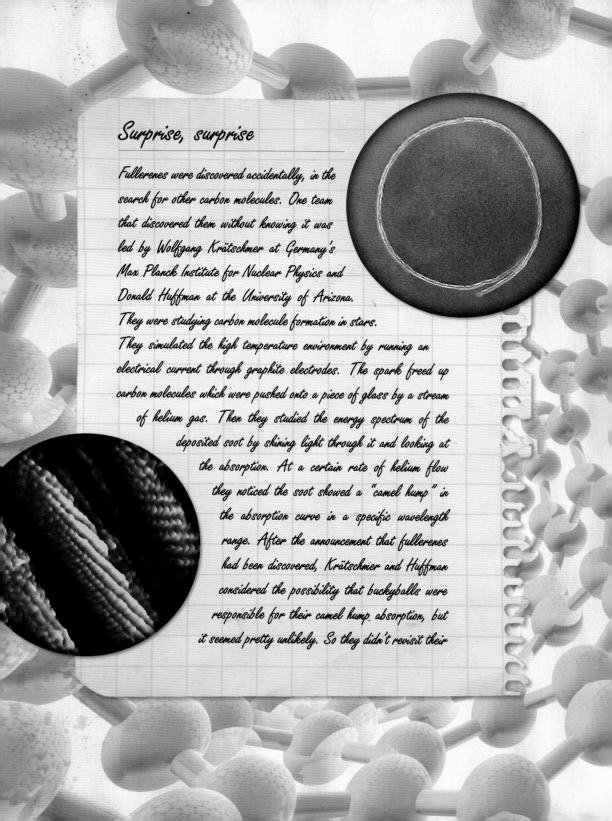

Surprise, surprise

Fullerenes were discovered accidentally, in the search for other carbon molecules. One team that discovered them without knowing it was led by Wolfgang Krätschmer at Germany's Max Planck Institute for Nuclear Physics and Donald Huffman at the University of Arizona. They were studying carbon molecule formation in stars. They simulated the high temperature environment by running an electrical current through graphite electrodes. The spark freed up carbon molecules which were pushed onto a piece of glass by a stream of helium gas. Then they studied the energy spectrum of the deposited soot by shining light through it and looking at the absorption. At a certain rate of helium flow they noticed the soot showed a "camel hump" in the absorption curve in a specific wavelength range. After the announcement that fullerenes had been discovered, Krätschmer and Huffman considered the possibility that buckyballs were responsible for their camel hump absorption, but it seemed pretty unlikely. So they didn't revisit their experiment until 1989. They then discovered that not only were they making buckyballs, they were making them more efficiently than anyone else could. This technique made enough for measuring and verifying the structural details that Smalley and Kroto had proposed.

In 1991, Sumio Iijima of Meijo University in Japan was duplicating their technique when he made a chance discovery of his own. Rather than looking just at the soot on the glass, Iijima looked at the surface of the graphite electrodes. There he found an entirely new form of carbon molecule: long thin tubes. Iijima had found what we now know of as carbon nanotubes. Carbon nanotubes are already being used as compact X-ray generators, television electrodes, transistors, and membranes for fuel cells.

It's quite possible that Iijima's chance discovery will fuel more applications than buckyballs themselves.

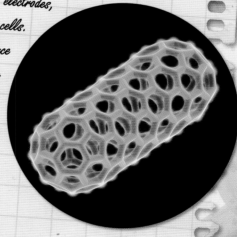

ALEX MÜLLER AND
GEORG BEDNORZ

DISCOVER HIGH-TEMPERATURE
SUPERCONDUCTIVITY

1986

RUNNING COLD, RUNNING FREE

OUR CIVILIZATION *runs on electricity. Electricity provides entertainment, comfort, and the tools of productivity at home, at school, and in our factories. But just about everywhere electrical current runs there is electrical resistance, which means heat and waste. Alex Müller and Georg Bednorz didn't start out working on that problem, but accident and coincidence may have put them in a position to solve it.*

Electricity is used by people everywhere at work and in the home—but electrical resistance from poor conductors means waste and generation of excess heat.

MORE COULD BE DONE WITH LESS

In 1911 a Dutch scientist by the name of Heike Kamerlingh Onnes built a cryogenics laboratory where he could lower the temperature to 1K, the equivalent of -458°F (-272°C). At that temperature, Kamerlingh Onnes found the electrical resistance of some metals would disappear entirely. That is, electrons in one of these metals, which he eventually called "superconductors," would move so smoothly there would be no wasted energy—no losses, no heat. The prospect of electrical current with no resistance excited both academic scientists and industrial engineers. In electrical power generation, for example, wires rotate through a magnetic field to produce lots of current. Even a little resistance heats the wires, reducing efficiency. If Kamerlingh Onnes's superconductivity could eliminate resistance, more could be done with less.

But a couple problems surfaced. First, super-conductivity turned out to be a pretty delicate state. If a superconductor was put in a magnetic field, the superconductivity would go away. The second problem is that superconductivity only set in at very low temperatures, and would disappear if the temperature got too high. Over the decades, new materials were developed that maintained their superconductivity up to 23K, or -418°F (-250°C). The only way to reach such temperatures was with liquid helium—an unwieldy, expensive proposition.

Heike Kamerlingh Onnes, Dutch physicist, pioneered refrigeration techniques which led to the discovery of superconductivity.

Kamerlingh Onnes gathered experimental evidence for the atomic theory of matter.

RIGHT: *Liquid helium is used as a coolant for maintaining the superconductivity of different materials but apparatus for using it is both unwieldy and expensive.*

215

CLOCKWISE FROM LEFT: *John Bardeen, Leon Cooper, and John Schrieffer, who discovered that electrons could be paired up at very low temperatures.*

JUST TOO IMPRACTICAL TO BE OF USE

Both of those problems were manifestations of a deeper issue: nobody knew how superconductivity worked. That changed in 1957. John Bardeen, Leon Neil Cooper, and John Robert Schrieffer determined that at very low temperatures, electrons were paired up. In those so-called "Cooper pairs," the electrons didn't bump into the atoms in the metal, but the two paired electrons sailed through as if the other atoms weren't there. The problem is that as the temperature rises to about 30K, the pairs are broken and superconductivity can't exist.

So over the decades from 1911 to 1986, superconductivity changed from a scientific hotbed and potentially game-changing industrial technology into a field with no scientific interest and limited industrial applications. Not that the field was without changes. Aside from the increase

in temperature to 23K, scientists discovered new superconductors that could tolerate high magnetic fields, and there were even some ceramic materials that demonstrated superconductivity at very low temperatures. But there was nowhere to go. Liquid helium temperatures were just too impractical to be of any general use.

Superconducting ceramics lose any electrical resistance when cooled to their individual Tc (critical temperature).

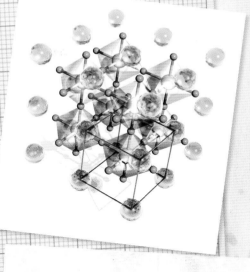

An entire new class of superconducting ceramics is based on the perovskite crystal structure.

THROUGH A VIBRATION IN A CRYSTAL

By 1980, Swiss physicist Alex Müller had studied perovskites, a particular class of crystalline ceramics, for 15 years. He worked at the IBM Zürich Research Laboratory in Rüschlikon, Switzerland, studying the interaction between electrical and magnetic effects within perovskite crystals with different compositions. His colleagues at IBM's much larger research laboratory in Yorktown Heights, New York, asked him to work with them for a couple years. Müller's visit just happened to coincide with the time they were working on making a superconducting computer. Müller worked with them for about a year and a half before the project was abandoned, and he returned to Switzerland. He returned home with a new fascination: superconductivity.

From everything known about it at that time, superconductivity worked by connecting two electrons to each other instead of to the atoms of the material they came from. It's hard to understand this mechanism without getting deep into details, but the electrons in a superconductor are kind of "pushed along" because they're linked to vibrations in the crystal that don't "drag" on the electrons. The key point is that electrons must be paired up for this mechanism to work, and they can only stay paired up to a temperature of about 30K. The highest temperature that had ever been reached was 23K.

Müller thought he could do better.

He planned to take advantage of another quantum mechanical complication called the Jahn-Teller (JT) effect. The JT effect causes a "stretching" in certain crystal structures. The stretching doesn't require any force pushing atoms away from each other, the rules of quantum mechanics simply don't let the electrons in one atom be too close to the electrons in another atom. That is,

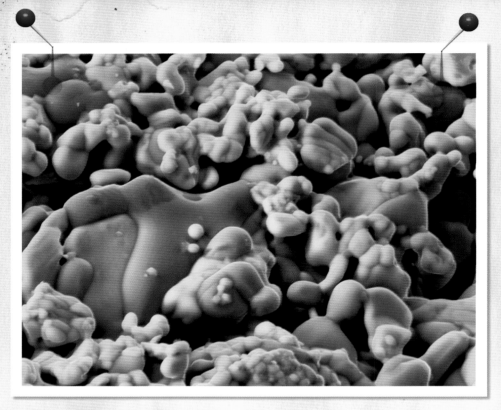

An SEM of a ceramic made of a superconducting alloy.

the electrons are coupled through a vibration in the crystal. Müller thought the JT coupling could possibly provide a stronger mechanism for superconductivity, a mechanism that would still work at higher temperatures.

PUTTING IN LONG AND UNUSUAL HOURS

So in 1983, back in Rüschlikon, he sought out help from German physicist Georg Bednorz. In his previous work, Bednorz had developed his own fascination with superconductivity; so he leaped at the chance to work with Müller, even though it meant putting in long and unusual hours. By this time Müller was an IBM Fellow, which meant he could work on pretty much whatever he wanted to, but Bednorz needed to work on other projects; so much of his work with Müller was after hours, on borrowed equipment.

It must have been a frustrating situation, made even more frustrating by their lack of progress. From 1983 to 1985 Müller and Bednorz made various ceramic compounds with strong JT effects and none of them exhibited superconductivity. If not for two chance events, they probably would have given up.

First, the administration at the laboratory decided that Müller and Bednorz's search deserved some support. They were given official access to automated measuring equipment, meaning Bednorz could work more normal hours. But still they had no success creating any superconducting material. Their lack of success led directly to the next chance intervention. While taking a break from the project, a deliberate attempt to create some "thinking space," Bednorz read a paper about a new material. A French research group had found that a barium-lanthanum-copper-oxygen (Ba-La-Cu-O) ceramic could sometimes act like a metal, that is, it could be made to conduct electricity at normal temperatures. That didn't mean it would work as a superconductor, but, on the other hand, it might exhibit a strong JT effect.

Müller and Bednorz tried it.

MORTAR AND PESTLE CHEMISTRY

This is literally mortar and pestle chemistry. Bednorz ground, mixed, and baked different powders. He ended up with a brittle solid. Bednorz took the sample and lowered its temperature. At 11K the resistivity dropped precipitously. Müller and Bednorz played with other mixtures, varying the ratio of Ba to La to modify the electronic structure of the final ceramic. They found the resistance drop at temperatures as high as 35K! This was a good sign they had made a superconducting material, but to be certain they would need to verify that the material "pushed out" a magnetic field, something known as the Meissner-Ochsenfeld effect, which is a universally accepted test for superconductivity. They didn't have the equipment to make that measurement. They ordered it, and while they were waiting they published a paper called "Possible High Tc Superconductivity in the Ba-La-Cu-O-System." By the time the paper was published they had made the measurement and verified they had set a new record for the highest temperature superconductor ever made.

But it was far more than that. They had proven that our understanding of the superconducting effect was incomplete. The old rules had said nothing could superconduct above 30K. They had demonstrated it was possible. And if one material could do it, then surely there must be more. In short months other groups demonstrated superconductivity at higher temperatures —even temperatures above 77K, which meant those materials could be used with liquid nitrogen cooling, a much more practical operating temperature.

A WHIRLWIND OF SCIENTIFIC INVESTIGATION

With one experiment, Bednorz and Müller had resurrected a moribund field and provided new hope for significant technological innovation. Their discovery stimulated a whirlwind of scientific investigation, but the one thing that has not yet arrived is an explanation. The only thing almost certain is that it is not explained by the Jahn-Teller effect! Chance had already intervened in Müller's work at Yorktown Heights, in finding Bednorz perfectly prepared to investigate superconductivity, in the timing of the French group's paper. But the biggest chance has yet to be explained: Müller and Bednorz discovered new materials that superconduct at high temperatures, but no one can explain the superconductivity.

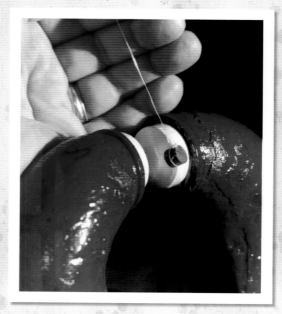

The Meissner effect is the expulsion of a magnetic field from a superconductor during its transition to the superconducting state.

Exactly the same, except for the music

Bednorz and Müller published their results almost stealthily. They didn't want to claim they'd succeeded before they could be sure, but they also didn't want to lose their claim for making the discovery. So their initial paper almost went unnoticed. Almost.

Three groups had noticed the relatively obscure paper and, by the end of 1986, they had duplicated Müller and Bednorz's work, then extended it to new materials at higher temperatures.

The word spread. These materials were so easy to make, anyone with a lab bench and an oven could try to make a new compound. In March of 1987 at the annual meeting of the American Physical Society, organizers convened a special conference on high temperature superconductors. Instead of the few hundred who might normally attend such a session, more than 3,000 tried to pack into a 1,200-seat room. The session, set to take place during an open spot in the schedule

Many new superconducting materials have been discovered, but they are difficult to work with; so putting them in practical forms—like this wire—is still problematic.

(7pm on a Wednesday evening), had to fit in scores of speakers, most of them limited to five-minute talks. Even so, the session went to 3:15 the next morning, with speaker after speaker interrupted by applause. More than one attendee said it was "the Woodstock of physics." The scientists felt their world had changed.

Since 1987 more high-temperature superconductors have been discovered, pushing the limit as high as 125K, well above the liquid nitrogen temperature. But these materials are difficult to manufacture, difficult to control. The balance of the materials is so delicate that one part of a sample can be superconducting while another isn't, and making such things as wires or sheets is problematic. And there is still no theoretical understanding of high-temperature superconductivity. But the promise of the field is still alive. When the potential is realized, superconductivity will indeed change the world.

A section through a superconducting cable. The balance of the materials is so delicate that one part of a sample can be superconducting while another isn't.

TIM BERNERS-LEE

CREATES THE WORLD
WIDE WEB

1990

SPINNING A WEB

IN 2009 NEARLY TWO BILLION PEOPLE

used the World Wide Web to access documents, images, videos, music and more. That information was stored on hundreds of thousands of World Wide Web servers all around the world. Back in 1990 there were 40 million personal computers worldwide, each with files like documents, images, videos, music, and more. But in 1990 there was one Web server. That single server seeded what may be the most significant revolution in human history. If not for a single man's frustration with organizing his own computer files, the human interconnections now commonplace through the World Wide Web would be completely different.

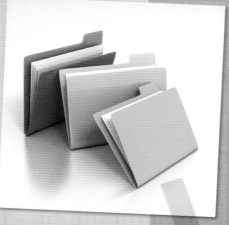

The World Wide Web has revolutionized the way that documents and data are shared and retrieved.

Tim Berners-Lee upgraded the concept of hypertext, the Internet language used to facilitate sharing and updating of information among researchers.

SUBATOMIC PARTICLES

In 1980 Tim Berners-Lee, a few years after receiving his degree in Physics from Oxford University, did some consulting work at CERN. CERN was founded as the Conseil Européen pour la Recherche Nucléaire, the European Council for Nuclear Research, but now the organization is known as the European Laboratory for Particle Physics. CERN's primary facility is an accelerator complex where subatomic particles are sped up with electric fields and kept on track with magnetic fields. In 2010 the Large Hadron Collider (LHC)—a ring nearly 17 miles in circumference—was brought online. But way back in the early planning

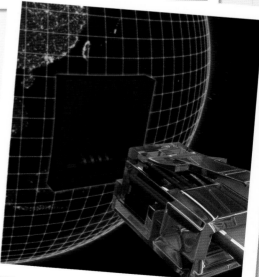

The Internet has changed the way we interact with each other as a global community, and has revolutionized the sharing of data.

stages of the LHC, Berners-Lee was hired to help with software development.

The problem was that CERN is a big place with multiple projects and many people, and Berners-Lee found himself getting buried by a mountain of information. He was assigned to help develop software for different projects, but he couldn't keep track of such things as the people, the project goals, the hardware requirements, and the schedule. Just to help keep his mind unscrambled, he wrote a program that allowed him to link data of different sorts: documents, requirements, contact information, drawings—just about everything he could need. Berners-Lee's program was a kind of hypertext notebook that was structured in such a way as to keep track of the flexible connections between projects, people, groups, experiments, software modules, and hardware devices. Within his program, all sorts of data could be linked in different ways, parallel to the way the brain flexibly organizes information. Because his program allowed him to organize just about anything, he named it after a popular Victorian handbook that provided useful advice on just about everything, *Enquire Within Upon Everything*. He called his program "Enquire."

IT STARTED WITH A MESS

Berners-Lee finished his work with CERN and didn't return until 1984. When he came back, the LHC planning was well underway, and CERN was getting more complicated. The LHC was much bigger than the existing particle accelerator. It had more detectors, and they would produce more data. The LHC required international collaboration at every phase. Berners-Lee was assigned to help with the management of that information. As he looked at the problem, he realized that the information management issues faced by CERN and the international team had parallels to the issues he personally faced organizing his own information.

In 1989, Berners-Lee proposed to take the program he had developed to help him keep track of his own work and modify it to optimize the sharing of information among the scientists and engineers collaborating on the LHC. A CERN systems engineer, Robert Cailliau, helped Berners-Lee get CERN behind the project. By 1990, Berners-Lee had realized the potential of his creation. He had written a program to allow him to access information on his computer with the same kind of associations that he could use within his own mind.

To be of any use to CERN, the program needed to provide access to information to users around the world no matter what computer they were using. Berners-Lee upgraded HTML—the hypertext markup language—adding more

Berners-Lee wrote a program just to help him keep track of the people, documents, requirements, contact information, and drawings needed on his project.

flexibility and capability. He defined the hypertext transfer protocol, which is more commonly seen in its acronym form, http (perhaps even more recognizable as "http://"). He even defined what we now know of as the URL, or Web address. As Berners-Lee thought about the capabilities he had developed and the requirements for connecting users from around the world, he realized he was creating much more than a flexible platform for allowing particle physicists to collaborate. The Internet already existed. That is, there were physical connections between computers spread across the continents, and there was a communications protocol that allowed computers to understand each other. Berners-Lee recognized that the program he had first developed for his own use, then adapted for use by CERN, had the potential to become an information-sharing tool of unprecedented power. He demonstrated his understanding with the name. The program would no longer be called "Enquire." He called his upgraded program "WorldWideWeb."

MORE THAN SHEER SIZE

In 1990 the first Web server, info.cern.ch, came online. Berners-Lee and Cailliau pushed CERN not only to develop WorldWideWeb (WWW) for internal use, but to spread the word beyond the confines of CERN. To Berners-Lee, the value of the WWW was clear, but others needed convincing. The first server in the United States began operating at the Stanford Linear Accelerator (SLAC) in December 1991. A year later there were 26 servers. A year after that, 200. By the end of 1994, 2,000 servers were sending the equivalent of the complete works of Shakespeare every second. In 2010 the same amount of data was being sent every millionth of a second.

More than the sheer size of the Internet, the World Wide Web has changed the way in which we do our work, the way in which we play, even the way in which we interact with friends and family. Tim Berners-Lee set out to create a program that organized his computer data so that he didn't have to struggle to keep his brain organized. By chance he had developed exactly the right tool to create interconnections between ideas and between people on a global scale. He had the right preparation to recognize what his program was capable of, and his desire brought his creation to the world.

The Internet is now used by people all over the world to communicate with each other daily.

The nature of the beast

Tim Berners-Lee created the World Wide Web, but he did far more than just release a new type of computer program into the world. He created a vision for how his creation would—and should—be used. He demonstrated his commitment to that vision from the beginning. Berners-Lee saw the Web as a resource for connecting people with people, and people with information. He could have patented his program and the protocols he developed, but that would have driven other organizations to make their own protocols. The Internet would have become like a superhighway system, but certain exits would only be open to particular users. Berners-Lee made sure that it wouldn't matter what hardware or software platform a user was on, they'd still have access to everything. He also pushed CERN to make the Web royalty-free.

Even more, Berners-Lee left CERN in 1994 to head the World Wide Web Consortium (W3C), an international organization that sets standards to push the Web toward living up to its potential. Berners-Lee's initial work has paid off in large part because he didn't make assumptions about the form of the data that would be presented and sought after. That philosophy is infused into the organization. The W3C develops and maintains open standards that lead to open development. Rather than having hundreds of companies developing different ways to define the color of an object, for example, the companies work with the same definitions. The companies then distinguish themselves by, say, how rapidly a browser can interpret and display the color information within a file. Just as in Berners-Lee's vision, with standards, even when companies are competing, they're collaborating to improve the Web.

Even computer and software companies in competition with one another are, whether they like it or not, collaborating to improve the World Wide Web.

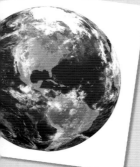

JUST SPINNING THE WHEELS

SOME SCIENTIFIC RESEARCH *is almost spur of the moment. It only takes a second to grab some chemicals and mix them, perhaps a couple weeks to change mirrors and lenses to reconfigure a laser experiment, maybe a year to plan and schedule time at an astronomical observatory. Some scientific research requires years of advance planning. For example, doing particle physics experiments at an accelerator can require years of lead time. But in the world of scientific research, one class of experiments stands out as requiring the most detailed planning furthest in advance: experiments done in the remote reaches of outer space. But even in this most rigorously planned branch of science, there is still room to react to chance observations. In February of 2005, an accidental discovery by the Mars rover Spirit led to some compelling evidence that free-flowing water was prevalent in the past of the planet Mars.*

Composite image showing Mars Opportunity Rover on the rim of Victoria crater. This is one of two identical rovers sent to Mars in the summer of 2003, their mission being to analyze soil and rocks and search for evidence of water.

THEY CAN'T GRAB A PLANET

Geology is the study of the origin, structure, and evolution of the Earth. Geologists have a tough job. They can't grab a planet, put it on a bench, and do some experiments. They make field observations and try to fit those observations into a theory of planetary formation and evolution. Each structure they measure got there as a result of two influences: 1) the general laws and principles of geology, and 2) the specific circumstances at that location. To reach a general conclusion, geologists have to identify and eliminate the contribution of the specific circumstances. The only way to do that is to make many observations in different locations. But no matter how far they travel, they are limited to one specific location: the Earth. And they're also limited in one more way: they can only observe things that have already happened. To predict what will happen in the future, geologists must uncover the general principles governing such things as mineral formation, erosion, and the interaction between solar energy and atmospheric dynamics. To discover the general principles—and then be confident in those conclusions—geologists need to gather as much data as possible. Perhaps they can't make another planet, but they can visit one.

ROOM TO REACT TO CHANCE OBSERVATIONS

Of all other planets in the solar system, Mars is the most similar to Earth. The Mars Exploration Program of the United States' National Aeronautics and Space Administration (NASA) has sent about a dozen spacecraft to the red planet. In practical terms, a spacecraft is about as far from a laboratory bench as it's possible to be. A scientist can't say, "Oh, I forgot to get the sodium hydroxide" and reach over to the bench next door. A science spacecraft is a miniature laboratory, but it's a laboratory that needs all its instruments designed years in advance. The design process starts with an idea for an experiment, then the requirements are defined, then different approaches to meeting those requirements are evaluated, then finally the detailed instrument design can be done. Fabrication, test, and integration into the space vehicle takes more years. Then the spacecraft is launched, spends months or years getting to a destination, and is turned on and tested in place. Only then do the instruments start doing their work.

In January of 2004, NASA launched two rovers—small dune-buggy-like vehicles festooned with scientific instruments. The mission of the rovers, named "Spirit" and "Opportunity," was

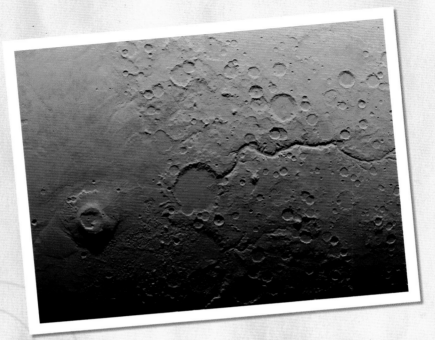

Gusev crater, Mars, is thought to have been formed three to four billion years ago by an asteroid impact. It is believed that water formed a river that drained into the crater.

to search for information about the history of water on Mars, as recorded in the geology of the red planet (or, more properly, the areology). Each rover was designed to last three months and to travel up to half a mile from its landing site. The rovers were more than 31 million miles from the scientists and engineers who operated them. At every step of the design and planning, dozens, perhaps even hundreds, of scientists and engineers had contributed.

THE OTHER SIDE OF MARS

The rover Opportunity landed in Eagle Crater and almost immediately discovered sedimentary rock, a clear indication that water had been present on the Martian surface for a long time. But on the other side of Mars, at a location called the Gusev Crater, Spirit had not found any minerals suggestive of widespread water. This was particularly puzzling

because the Gusev Crater is located right at the opening of what appears to be an ancient dry riverbed. Over the next three months, Spirit found nothing but dust, sand, and volcanic rocks. Spirit had put in its three months, and the mission could have been over. But the rover still had some legs; so NASA ponied up the cash to keep the mission control team together.

But mission managers were aware that they were on borrowed time; the rovers could fail at any moment. Now was a time for even more detailed planning, with even more attention to the value of time. They directed Spirit to the Columbia Hills, a region with rocky rises named after astronauts who had died in the Columbia space shuttle re-entry disaster. They sent Spirit to the top of Husband Hill, but the rover couldn't get traction to move up the slope. The wheels churned up the Martian soil as the rover made its desperate bid to reach the top. Each moment spent here represented a delay on the way to the scenic viewpoint that would help

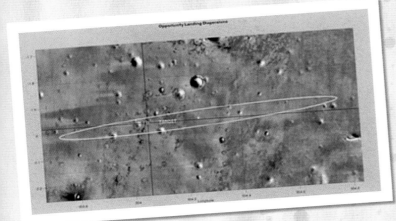

ABOVE: *Columbia Hills, a range of hills inside Gusev Crater.*

LEFT: *A close-up of Opportunity's predicted landing site at Meridiani Planum. The region was suspected to contain gray hematite, an iron-rich mineral that usually forms in the presence of water on Earth.*

The Instrument Deployment Device, or the "arm" from the Mars Rover Spirit, extending to the "Adirondack" rock on the surface of Mars.

the mission scientists decide where to send the rover for what might be its last precious days. They backed Spirit, readying the rover to try a different path up the hill.

When mission planners looked at images from the rover so they could choose a new path, they noticed Spirit's wheels had churned up a bright white, sandy soil different from anything they'd yet seen. They decided to take a chance. Rather than continue with their meticulously developed plan, they directed the rover to gather reflected X-ray data from the white soil. Then they sent the rover back up the top of the hill. The rover's wheels still slipped, churning up more white soil. When the X-ray data from the previous day was analyzed, the results were very intriguing. Since Spirit had churned up more white soil, they took more data right there.

The white soil was composed of a mineral salt, a chemical whose presence strongly indicates that the area of the Columbia Hills was once covered in water. Spirit's chance discovery—and the mission team's decision to take advantage of the accidental observation—confirmed that signs of surface water on Mars were not limited to one location. The mission team allowed their desire to understand outweigh their need to meet a mission plan, and because of that they made an accidental discovery with implications for two worlds.

Following a long drive, NASA's Mars Exploration Rover Spirit took this backward glance at its tracks across the landscape 90 Earth-days into its mission. The image is from Spirit's navigation camera.

A CITY ON MARS

Often, as we gaze into the night sky, we wonder if the other planets that circle our sun are populated as is our own world. Are there cities on them? If so, what are they like? Artist Paul here depicts an imaginative, yet scientifically accurate conception of a typical city on Mars. (Details on page 144)

Popular images of Mars and attempts to explore it. There has always been speculation about the possibility of escape from Earth, and life on other planets. With the discovery of water on Mars, the public's imagination was reignited.

Live long and prosper

Every space mission has performance objectives——a set of goals that must be met to consider the mission successful. Scientists and engineers design a spacecraft and plan its operations to meet those goals. But just because a program meets its goals in a year, for example, doesn't mean it's no longer useful. Spirit and Opportunity were designed to travel about a half a mile in their three-month missions. More than five years later, the two rovers were still going. Spirit, hampered by a malfunctioning wheel, had logged more than four miles of travel, while Opportunity had more than 12 miles on the odometer.

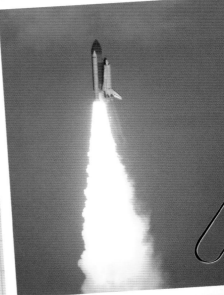

As impressive as that is, there are other space missions that have extended their missions even more dramatically. The Voyager 1 and 2 spacecraft were launched in 1977 with a four-year mission. They are now in their 33rd year of interplanetary travel, on their way well beyond the solar system. In 2009, data from both Voyager spacecraft revealed an unexpected magnetic field that shields our solar system from becoming immersed in the local interstellar cloud, a tenuous cloud of hydrogen and helium atoms. The Mars rovers racked up quite a few miles on the planet's surface, but the Voyager spacecraft have each traveled more than eight billion miles!

CHAPTER

LIVING WITH CHANCE: PREPARATION AND DESIRE

★

Looking at the role of chance in previous scientific discoveries, what lessons can we apply to ensure we take advantage of luck to enhance future progress? To answer that question, we first need to understand where and how chance appears in scientific discovery. There are plenty of historical and philosophical perspectives on the nature of scientific progress, but they agree that knowledge grows in two ways: slowly, steadily, incrementally; or through flashes of insight, sudden realizations that things are not the way they have seemed.

This second mechanism—the flash of insight, the sudden discovery—can come about as part of a planned research process, or it can come by accident. How much scientific progress is by accident? It's debatable, but what can't be argued is that chance has had a role in scientific discovery, and, if we're smart, we'll make certain it continues to play a role in the future.

We have seen that preparation, opportunity, and desire all have to be present for an accidental discovery to come to fruition. To encourage future accidental discoveries, the structures of science must create an environment that facilitates preparation, provides opportunity, and nurtures desire. The anecdotes presented here may show the way.

SCIENTIFIC PROGRESS

Science is a process for learning about our world. Scientists make observations, come up with a mental model that explains those observations, and test their model. In scientific nomenclature, the mental model is called a theory. Scientists spend much of their time making predictions to test the edge of the theory.

To give a silly example, suppose that I have observed that chewing gum tastes better in the first few moments after I put it in my mouth. I develop a theory that foods taste better when they're first eaten, and I test this theory with one chewing gum after another. The experiments support my theory! Then I decide to push the edges of the theory by tasting an apple. Then some mashed potatoes. Finally, a glass of wine. With each new test the result is a little less certain until the test with the wine: it gets better several seconds after I taste it. My theory collapses. I need a new one, one that explains why gum tastes better at first and wine tastes better later.

American astronomer Carl Sagan popularized astronomy and astrophysics.

Much of scientific progress is like that, a planned series of tests that support or disprove a theory. If the theory is very strong—that is, it is supported by lots of experimental results—then it takes some pretty convincing tests to show the theory is wrong. The scientific community will resist changes to the theory. As Carl Sagan said, "extraordinary claims require extraordinary evidence." When an existing theory is disproven, a leap of insight is needed to create a new theory.

In this description of scientific progress, the process seems straightforward and logical, with each step part of a plan. But, as astrophysicist Amadeo Balbi wrote:

> *Unfortunately, those who follow the scientific enterprise from within know very well that things to not always go exactly this way. In their chess game against nature, only rarely do scientists have a clear vision of the next move.*[1]

This is where chance steps in. Sometimes chance intervenes by creating circumstances that put someone in a situation that encourages new theoretical insight, but far more often an accidental observation provides unexpected evidence outside of an existing theory. The accidental observation may require an existing theory to change, it might cause an existing theory to topple, or it can even open up a new field of inquiry.

Wallace Carothers and his explorations of polymer creation pushed the boundaries of chemistry and made us change our understanding of molecular assembly. Friedrich

1 Balbi, A. (2008). *The Music of the Big Bang*. Berlin Heidelberg: Springer-Verlag.

Wöhler's laboratory synthesis of an organic molecule (eventually) destroyed the theoretical framework of the Vital Force Theory. Karl Jansky's accidental discovery of radio signals from space created the field of radio astronomy. There's good evidence that chance observations sometimes provide just the right push to move science ahead; so it's worth looking at the different types of chance, to see what can be encouraged.[2]

THE FLAVORS OF CHANCE

Chance appears in scientific discovery in different flavors:

- Right-place, right-time. Random convergence of circumstances creates a situation that just happens to lead to a discovery.
- Exhausting possibilities. Chance is "manufactured" when a researcher just blindly tries every different conceivable approach to solve a problem.
- Coincidental circumstances. A focussed research plan works for the wrong reasons, or gets the sought-after result in an unexpected fashion.
- Serendipity. A well-planned investigation turns up an unexpected and unrelated observation, which is then followed to its conclusion.

Each of the accidental discoveries described in this book fits into at least one of the categories described above.

Anytime we look back on an event—in our personal history or in the world's history—we see the convergence of chance circumstance necessary for the event to have taken place. If you hadn't given in to your craving for a candy bar at exactly that moment you wouldn't have bumped into your friend who invited you to the movie where you met your fiancé. In this sense, everything in our lives is the result of chance—events just happened to unfold in a specific way. This "right-place, right-time" variety of chance is exemplified by Abu Ali al-Hasan Ibn al-Haitham and Isaac Newton, whose random circumstances put them in a situation that increased the likelihood they would make their theoretical breakthroughs. Different circumstances may have led them to those same breakthroughs, just in a different manner or at a different time. Tim Berners-Lee also just happened to feel scatter-brained enough

Newton pondered the effects of gravity in his garden.

2 We're just going to touch on the topic. For readers interested in a more scholarly investigation of the different types of chance involved in scientific discovery, take a look at one perspective in Austin, J. H. (2003). *Chase, Chance, and Creativity*. MIT Press.

ABOVE AND RIGHT: *Inventor Thomas Edison in his workshop in West Orange, New Jersey.*

to want a computer to help him keep track of multiple responsibilities, and then he was insightful enough to see the connection between the tool he had developed and a need at the place where he just happened to hold a job. Sooner or later someone would have developed something equivalent to the "WorldWideWeb" program. Luckily for humanity, it happened to be a person with Berners-Lee's attitude toward free and open sharing of information. These random associations of person and place are just that: random.

The "exhausting possibilities" variety of chance is almost inevitable. Thomas Edison, one of history's great inventors, does not appear in the pages of this book because he almost eliminated chance in his developments.[3] For example, depending upon which source you believe, Edison tested somewhere between 6,000 and 10,000 different materials while looking for an appropriate filament for his incandescent lamp. Sooner or later he was bound to find something that worked! Charles Goodyear took advantage of chance in the same way: trying every darn

Thomas Edison listens to the first phonogram sent to New York from England.

---

3 One of Edison's assistants, William J. Hammer, did accidentally invent the electron tube—the invention at the heart of radio and television for more than half a century—but neither Edison nor Hammer had the desire to bring the chance observation to fruition.

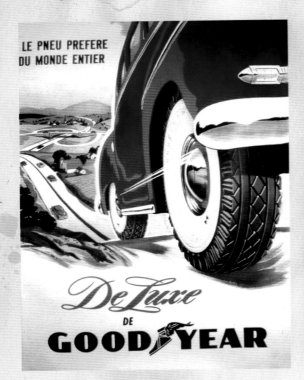

LE PNEU PREFERE
DU MONDE ENTIER

DeLuxe
DE
GOOD YEAR

GOOD YEAR
STRAIGHT SIDE TYRES

The Goodyear Tire and Rubber Company was founded in 1898, 30 years after Goodyear himself died penniless. It was, however, great timing. The bicycle craze of the 1890s was booming. The horseless carriage, which some ventured to call the automobile, was a wide-open challenge.

thing he could think of to reach a result. But this approach is not as effective as one might think: even with such a "try everything" approach, Goodyear still needed a chance event to invent vulcanized rubber, and Edison's solutions were not always the best. For example, although the filament he selected proved electric lighting was feasible, it was soon replaced by more robust materials. The shotgun Edison approach will often result in something that's "good enough,"[4] but rarely leads to a significant breakthrough.

That's not to say that focussed research is unproductive. On the contrary, it's the most productive endeavor in scientific and technological programs. To push scientific discovery, experiments are designed at the edge of theories, and the theory is supported or contradicted by the experimental result. Scientists go into these experiments with an idea, a prediction of the result. But even if the result doesn't match the prediction, scientists still expect something close to the prediction. Here's where "coincidental circumstances" chance comes in. It would have been easy for Stephanie Kwolek to chuck out the cloudy liquid that became DuPont™ Kevlar®, or for Oersted to disregard the

---

4 In the aerospace industry, for example, this approach is often used. It works, but only if you accept the industry mantra, "better is the enemy of good enough." That phrase haunts the nightmares of scientific and technical personnel, who are always searching for "better."

Oersted demonstrates his ground-breaking discovery of electromagnetism (magnetism produced by an electric charge in motion) to colleagues.

small electromagnetic effect that didn't fit his expectation. Or look at the story of Ignaz Semmelweis (pictured right)—his results were so far from the expectations of his contemporaries that they were willing to sacrifice more lives rather than entertain the possibility Semmelweis was onto something.[5] To take advantage of this flavor of chance, researchers must be absolutely committed to following where the chance observation leads, rather than letting preconceptions blind them to possibility.

The final type of chance is called "serendipity." The word serendipity has now percolated through academic and popular expressions.[6] It was originally introduced in 1754 by Horace Walpole, in homage to adventures detailed in a 1557 Italian story called *The Three Princes of Serendip*. In the story, "through accident and sagacity" three royal brothers learn answers to questions they weren't asking. Walpole recognized that scientists in search of some specific result often find an answer to a question they weren't asking. William Henry Perkin

---

5   An embarrassing (although—thankfully—not as serious) example from my own career: a subsystem of an experiment designed for the Space Shuttle was failing. A colleague of mine noticed that the failure never occurred when two separate instruments were not connected. I couldn't imagine how that would make a difference; so it took three weeks to find out he had been right: an unexpected electrical interaction between the two instruments was causing the subsystem to fail.

6   For those interested in the history of the word itself, try Merton, R.K. and Barber, E. (2004). *The Travels and Adventures of Serendipity*. Princeton, NJ: Princeton University Press.

Horace Walpole was responsible for coining the term "serendipity."

was looking for antimalarial medication and he found fabric dye. Ross Harrison was trying to understand neural development and he invented in vitro tissue culturing. Percy Spencer was trying to develop aircraft radar and he revolutionized home cooking. Scientists who want to be receptive to a serendipitous discovery need more than an open mind; they require a broad enough knowledge base to be aware of the importance or implications of a discovery outside of their field of expertise.

MAKING YOUR OWN LUCK

Anyone reading this far in the book already knows what it takes to complete an accidental discovery: preparation, opportunity, and desire. One of the reasons scientists resist acknowledging the role of chance is that people outside of the sciences tend to think of a chance discovery as something that could happen to anyone. That is, if it was just us who were lucky enough to be in the right place at the right time, we could have been in Stockholm accepting that gold medal from the King of Sweden on behalf of the Nobel Committee.

Before we book our airplane tickets, we should think about how many people saw an apple fall to the ground before Newton was inspired to make the universe a place we could hope to understand. How many birthday candles have been blown out without anyone noticing the presence of buckyballs or carbon nanotubes? Penicillin, microwave ovens, X-rays—all had been seen by other people before they were discovered by scientists with the preparation to understand what they were seeing and the desire to learn the implications of the chance observation. If this were a book on unmade accidental discoveries, it would be a lot longer. Preparation and desire are essential parts of accidental discovery, and if we focus on maximizing "luck" without enhancing preparation and desire, any efforts to culture chance discovery will go for nought.

But, as these stories show, luck does play a role. So what needs to be done to cultivate

Sir William Henry Perkin was looking for a cure for malaria when he accidentally created dye.

Preparation, opportunity, and desire—the three ingredients necessary for an accidental discovery.

Chance has always had a part to play in scientific progress, and it always will.

chance interventions and stimulate scientific progress? To maximize the whole, maximize the parts: preparation, opportunity, and desire.

To maximize preparation, scientific teams must be composed of highly trained specialists who are willing to work together with colleagues from different disciplines. To maximize opportunity the team needs to actively investigate questions at the boundary of theory. Maximizing preparation and opportunity is straightforward. Maximizing desire, though, is a little more complicated.

Every scientist wants to expand the boundaries of knowledge. That's what it means to be a scientist. But today's world requires scientists to be goal-oriented, to make progress toward a well-defined target. Grants are awarded to scientists who have shown progress in the past. In the corporate or government research lab, scientists and engineers have a product to deliver. Anything that does not move toward the goal is a distraction and is discouraged.

Recently, the scientific community has become more aware of this problem, and there are now programs designed to provide funds for self-directed research for both promising and proven researchers. It's a necessary step to encourage accidental discovery, but it's also a limited step. Some accidental discoveries are just that: accidents. For every Charles Goodyear, who sacrificed so much to discover vulcanization, there are scores or hundreds who have spent their lives in a fruitless search for a stronger steel or a more robust fabric. Researchers who follow chance observations take the risk that they will find nothing of significance, and the agencies who fund them may eventually think their money was ill-spent. It will take some time to strike the right balance between goal-oriented and observation-driven research. As long as scientists and their supporters are aware there is a balance to be struck, there is hope that chance observations can be brought to fruition.

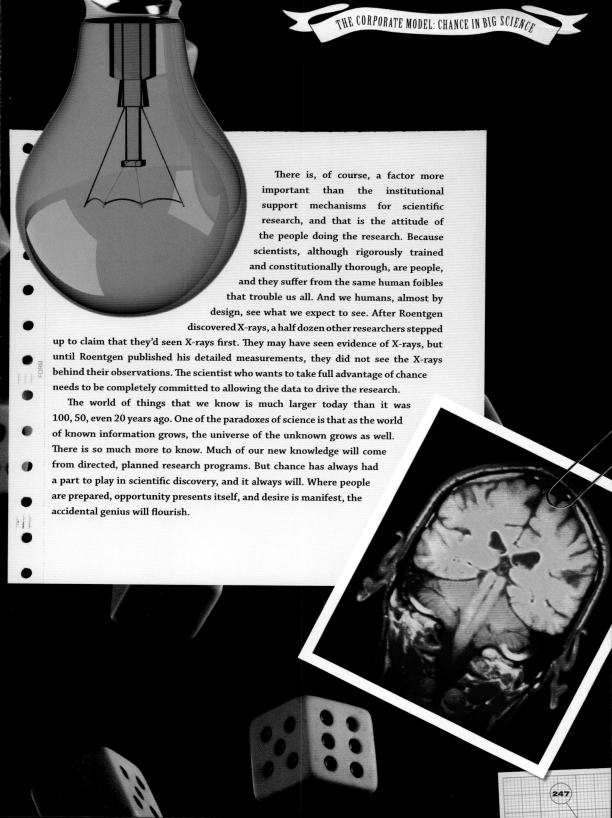

There is, of course, a factor more important than the institutional support mechanisms for scientific research, and that is the attitude of the people doing the research. Because scientists, although rigorously trained and constitutionally thorough, are people, and they suffer from the same human foibles that trouble us all. And we humans, almost by design, see what we expect to see. After Roentgen discovered X-rays, a half dozen other researchers stepped up to claim that they'd seen X-rays first. They may have seen evidence of X-rays, but until Roentgen published his detailed measurements, they did not see the X-rays behind their observations. The scientist who wants to take full advantage of chance needs to be completely committed to allowing the data to drive the research.

The world of things that we know is much larger today than it was 100, 50, even 20 years ago. One of the paradoxes of science is that as the world of known information grows, the universe of the unknown grows as well. There is so much more to know. Much of our new knowledge will come from directed, planned research programs. But chance has always had a part to play in scientific discovery, and it always will. Where people are prepared, opportunity presents itself, and desire is manifest, the accidental genius will flourish.

REFERENCES

★

General references that address, or at least touch upon, chance discoveries:

ACTON, J., ADAMS, T., & PACKER, M. (2006). *Origin of Everyday Things*. New York: Sterling Publishing.

AUSTIN, J. H. (2003). *Chase, Chance, and Creativity*. MIT Press.

BROWN, D. E., (2002). *Inventing Modern America: From the Microwave to the Mouse*. Cambridge, MA: MIT.

CARLISLE, R. (2004). *Scientific American Inventions and Discoveries: All the Milestones in Ingenuity—From the Discovery of Fire to the Invention of the Microwave Oven*. Hoboken, NJ: John Wiley and Sons.

FRIEDLAND, G. W. & FRIEDMAN, M., (1998). *Medicine's 10 Greatest Discoveries*. New Haven, Conn: Yale University Press.

KOHN, A., (1989). *Fortune or Failure: Missed Opportunities and Chance Discoveries*. Cambridge: Basil Blackwell.

MCGRAYNE, S. B. (2001). *Prometheans in the Lab: Chemistry and the Making of the Modern World*. New York: McGraw-Hill.

ROBERTS, R. M. (1989). *Serendipity: Accidental Discoveries in Science*. New York: John Wiley & Sons.

SHAPIRO, G., (1986). *A Skeleton in the Darkroom: Stories of Serendipity in Science*. San Francisco: Harper & Row.

Specific references, one for each anecdote. The reference list is obviously not comprehensive, but will give the interested reader a start on learning more about stories of interest.

BAKING SODA INTO GLASS
See page 22
An unnamed Phoenician, c.3000 BCE
Pliny the Elder. (Bostock, J., Trans.), (1855). *The Natural History*, XXXVI, 65. London. Taylor and Francis.

UNLOCKING THE SECRETS OF LIGHT
See page 26
Abu Ali al-Hasan Ibn al-Haitham reveals secrets of optics, c.1011
Sabra, A. (2003, September-October). "Ibn al-Haytham." *Harvard Magazine*, 106(1), pp. 54–55.

IT'S ALL DOWNHILL FROM HERE
See page 30
Isaac Newton discovers universal gravitation, 1666
Stukely, W. (1752). *Memoirs of Sir Isaac Newton's Life*. Ms. 142, The Royal Society Library, London. http://www.newtonproject.sussex.ac.uk/view/texts/diplomatic/OTHE00001 (Accessed online April 13, 2010).

A SPARK OF INSPIRATION
See page 36
Pieter van Musschenbroek and Ewald Jürgen von Kleist discover the capacitator, 1745–46
Mills, A. (2008, December). Studies in Electrostatics Part 6: The Leyden jar and Other Capacitors. *Bulletin of the Scientific Instrument Society 99*.

A Jump Sparks a New Understanding
See page 42
Luigi Galvani discovers "Animal Electricity," 1781
Bresadola, M. (2003). *At Play with Nature: Luigi Galvani's Experimental Approach to Muscular Physiology.* In F. L. Holmes, J. Renn, H-J. Rheinberger (eds.), *Reworking the Bench: Research Notebooks in the History of Science* (pp. 67–92). Great Britain: Kluwer Academic Publishers.

Seeing From a Different Angle
See page 48
Etienne-Louis Malus discovers light polarization by reflection, 1808
Arago, F. (1859). *Biographies of Distinguished Scientific Men* (W. H. Smyth, Rev. Baden-Powell, & R. Grant, Trans.). Boston: Ticknor and Fields, pp. 117–170. http://www.archive.org/stream/biographiesof02arag. (Accessed online April 13, 2010)

A Lecture on Current Affairs
See page 54
Hans Christian Oersted connects electricity and magnetism, 1820
Kipnis, N. (2005). *Chance in Science: The Discovery of Electromagnetism by H.C. Oersted.* Science & Education, 14: pp. 1–28

Bringing the Lab to Life
See page 60
Friedrich Wöhler, synthesizes organic molecules, 1828
Correspondence between Wöhler and Berzelius on Urea (n.d.). http://www.chem.yale.edu/~chem125/125/history99/4RadicalsTypes/UreaLetter1828.html (Accessed online April 13, 2010)

Fixing a Sticky Situation
See page 64
Charles Goodyear invents the rubber vulcanizing process, 1839
Slack, C. (2002). *Noble Obsession.* New York: Hyperion.

The First "Mister Clean" is not so Welcome
See page 68
Ignaz Semmelweis reduces the incidence of childbed fever, 1847
Semmelweis, I. (1861). *The Etiology, Concept, and Prophylaxis of Childbed Fever* (K. C. Carter, Trans. (1983)). Madison: University of Wisconsin Press.

A New World Starts with a Dye
See page 72
William Henry Perkin invents the first artificial dye, 1856
McGrayne, S. B. (2001). *Prometheans in the Lab: Chemistry and the Making of the Modern World.* New York: McGraw-Hill.

Sidestepping Fate
See page 78
Alfred Nobel invents dynamite, 1867
Kelly, J. (2006, Summer). *Big Bang: The Deadly Business of Inventing the Modern Explosives Industry.* Invention and Technology 22(1).

Chicken Soup Gone Bad—Good News for Health
See page 82
Louis Pasteur develops chicken cholera immunization, 1879
Fleming, A. (1947). Louis Pasteur. *The British Medical Journal*, 1(4502), pp. 517–522.

Jelling the Future of Biological Research
See page 88
Angelina Hesse develops a bacterial growth medium, 1881
Hesse, W. (1992). *Walther and Angelina Hesse-Early Contributors to Bacteriology* (D. H. M. Gröschel, Trans.). ASM News, 58(8), pp. 425–428.

SEEING SKELETONS
See page 94
Wilhelm Conrad Roentgen discovers X-rays, 1895
Shapiro, G. (1986). *A Skeleton in the Darkroom: Stories of Serendipity in Science*. San Francisco: Harper & Row.

TOO TOUGH TO BREAK
See page 102
Edouard Benedictus invents safety glass, 1903
Nova (1999, February 16). *Escape: Because Accidents Happen: Car Crash* [Television broadcast]. Public Broadcasting Service. Transcript available at http://www.pbs.org/wgbh/nova/transcripts/2605car.html (Accessed online April 13, 2010)

MAKING IT UP
See page 108
Leo Baekeland invents Bakelite, the first synthetic material, 1906
Anonymous, (2000). "Leo Hendrik Baekeland." *Contemporary Heroes and Heroines*, Book IV. Farmington Hills, MI: Gale Group.

TURNING THINGS INSIDE-OUT
See page 114
Ross Harrison develops tissue culturing, 1906
Friedland, G. W. & Friedman, M., (1998). *Medicine's 10 Greatest Discoveries*. New Haven, Conn: Yale University Press.

A MOLDY CHAIN OF CIRCUMSTANCE
See page 120
Alexander Fleming discovers penicillin, 1928
Macfarlane, G. (1984). *Alexander Fleming, The Man and the Myth*. London: Chatto and Windus.

BOUNCING AND RACING AROUND THE LAB
See page 126
Wallace Carothers and Arnold Collins, Synthetic Rubber, 1930
Wallace Carothers and Julian Hill, Nylon, 1930
McGrayne, S. B. (2001). *Prometheans in the Lab: Chemistry and the Making of the Modern World*. New York: McGraw-Hill.

ASTRAL VOICES ON THE TELEPHONE
See page 134
Karl Jansky opens the study of radio astronomy, 1931
Sullivan, W. (1981, November 17). "Radio astronomy, 50 years old, moves toward a new frontier." *The New York Times*. http://www.nytimes.com/1981/11/17/science/radio-astronomy-50-years-old-moves-toward-a-new-frontier.html (Accessed online April 13, 2010)

COLORING THE FIGHT AGAINST DISEASE
See page 140
Gerhard Domagk discovers the first antibacterial chemical, 1932
Wood, M. E. (2010, Spring). "Soldier Sulfa." *Chemical Heritage Magazine*, 28(1). http://www.chemheritage.org/pubs/magazine/mil_domagk.html (Accessed online April 13, 2010)

A BIG SLIP-UP
See page 146
Roy Plunkett invents DuPont™ Teflon®, 1938
E. I. du Pont de Nemours and Company. Roy Plunkett: 1938. http://www2.dupont.com/Heritage/en_US/1938_dupont/1938_indepth.html (Accessed online April 13, 2010)

A WARM POCKET LEADS TO A HOT COOKING TECHNIQUE
See page 154
Percy Spencer invents the microwave oven, 1945
Anonymous (1999, Jan/Feb). "Melted chocolate to microwave." *Technology Review*, 102(1). https://www.technologyreview.com/biomedicine/11834/ (Accessed online April 13, 2010)

ONE DROP DOES IT
See page 160
Harry Coover invents cyanoacrylate SuperGlue, 1951
Coover, H. W. (2000, September–October). "Discovery of Superglue shows power of pursuing the unexplained." *Research-Technology Management*, pp. 36–39.

A Slick Invention
See page 166

Patsy Sherman invents Scotchgard water repellent, 1953

Anonymous (2000). "Patsy Sherman." *Notable Women Scientists*. Farmington Hills, MI: Gale Group, 2000.

Piecing It Together
See page 170

Francis Crick and James Watson decipher the structure of DNA, 1953

Friedland, G. W. & Friedman, M., (1998). *Medicine's 10 Greatest Discoveries*. New Haven, Conn: Yale University Press.

Keeping the Beat
See page 180

Wilson Greatbatch invents the implantable pacemaker, 1958

Adams, J. (1999). "Making hearts beat." From The Smithsonian's Lemelson Center Invention Features: Wilson Greatbatch. http://invention.smithsonian.org/centerpieces/ilives/lecture09.html (Accessed online April 13, 2010)

Don't Gum Up the Works!
See page 186

Stephanie Kwolek invents DuPont™ Kevlar®, 1965

Brown, D. E., (2002). *Inventing Modern America: From the Microwave to the Mouse*. Cambridge, MA: MIT.

Receiving the Long-Distance Call
See page 192

S. Jocelyn Bell discovers pulsars, 1967

McNamara, G. (2008). *Clocks in the Sky: The Story of Pulsars*. Berlin; New York: Springer; Chichester, U.K.: Published in association with Praxis Publishing.

Plastics get Current
See page 198

Alan MacDiarmid, Hideki Shirakawa, and Alan Heeger invent intrinsically conducting polymer, 1977

Gorman, J. (2003, May 17). Plastic Electric. *Science News*, 163(20), pp. 312–313.

A Cinderella Story
See page 206

Harry Kroto, Richard Smalley, and Robert Curl discover fullerenes, 1985

Anonymous, (1992, January). "The search for carbon in space and the fullerene fallout." *Science Watch*, 3(1), 3-4. Reprinted in *Current Contents*, 37, pp. 3–7, September 13, 1993.

Running Cold, Running Free
See page 214

Alex Müller and Georg Bednorz discover High-Temperature Superconductivity, 1986

Schechter, B. (1989). *The Path of No Resistance*. New York: Simon & Schuster.

Spinning a Web
See page 222

Tim Berners-Lee creates the World Wide Web, 1990

Brown, D. E., (2002). *Inventing Modern America: From the Microwave to the Mouse*. Cambridge, MA: MIT.

Just Spinning the Wheels
See page 228

The Spirit Rover uncovers evidence of free-flowing water on Mars, 2005

National Aeronautics and Space Administration, Jet Propulsion Laboratory. 2/28/05, "Spirit taking in 'Tennessee Valley.'" www.jpl.nasa.gov/missions/mer/daily.cfm?date=2&year=2005 (Accessed online April 13, 2010)

CREDITS

★

All other images are the copyright of Quintet Publishing Ltd. While every effort has been made to credit all contributors, Quintet Publishing would like to apologize should there have been any omissions or errors—and would be pleased to make the appropriate corrections for future editions of the book.

T = top, B = bottom, L = left, R = right, C = center, BA = background

AIP Emilio Segre Visual Archives: IBM, W. F. Meggers Gallery of Nobel Laureates.

Alamy: 50 © Craig Eisenberg; 108B-L © Hugh Threlfall; 112 © Pictorial Press Ltd; 135C-R © World History Archive; 163T-R © UK Scenes; 163B-L © Phil Degginger; 164B-L © Chris Johnson; 194, 195T © Brian Seed.

AT&T: 14C-L Courtesy of AT&T Archives and History Center.

ChemHeritage: 109T © The Chemists' Club Collection, Courtesy of the Chemical Heritage Foundation Collections.

Corbis: 15 © Ross Swanborough/epa; 48T-C © Michael Nicholson; 78T-C, 95T © Bettmann 134C-T; 135B-R, 138T-L, 144 © Bettmann; 201B-R © Haruyoshi Yamaguchi/Sygma; 202 © Image Source; 206T-C © Paul Seheult; Eye Ubiquitous; 209T © Bettmann; 223T © Ed Quinn; 231B © NASA/JPL/ASU/CNP.

DuPont: 12T-R; 187T-R.

European Space Agency: 231T; 233B-R; 234BA.

Getty: 18, 41T-L © SSPL; 72T-C © Hulton Archive; 110R © SSPL; 145T-R; 149B © G. Wanner; 163T-L © Lalo Yasky/WireImage; 238 © Santi Visalli Inc.; 247B-R © UHB Trust.

Hagley Archives and Library: 126T-C, 131T-L, 132T-L, 146T-C, 186T-C © Courtesy Hagley Museum and Library.

istock: 8 © Jasmin Awad; 16 © Rischgitz/Getty Images; 19, 35B-L © Duncan Walker; 23T-R © 24B © Clint Spencer; 30T-L © Cecilia Bajic; 30C-L, 30B-L; 30B-R © simon mosse; 32T-L © nicholas belton; 48C-L © Louis du Mont; 51T © Jon Larson; 51B © Silke Dietze; 52 © Sergey Peterman; 54C-L © Lyle Koehnlein; 60C-L; 64T-C; 64C-L © Mayumi Terao; 64B-L © ray roper; 64B-R © Steven Wynn; 68C-L © Tim McAfee; 70C-R © Clifford Mueller; 74T © Linda Steward; 75B © Steven Wynn; 76B-L © nicoolay; 78B-R © Olena Druzhynina; 82C-L © Patricia Hofmeester; 82C-LB © Ken Brown; 83T-R © Sergey Mironov; 88C-L © Henrik Jonsson; 88B © Vasko Miokovic; 89T © James Driscoll; 91 © Alexander Raths; 93R © Ruud de Man; 98C-L ©

Lawrence Sawyer; 98B-L © Christopher Pattberg; 113 © Jeffrey Heyden-Kaye; 114T-L, 116C-L © Bruno SINNAH; 127 © Lorelyn Medina; 128B-L © Danny Smythe; 130 © Marek Uliasz; 133T-R © Dave Bluck; 133C © Sufi70; 137 © Vladimir Piskunov; 138BA, 139BA © Thomas Tuchan; 139R © Mike Norton; 140T-L, 140B-R © Ben Greer; 140C-L © luismmolina; 140B-L © Henrik Jonsson; 145T-L © Duncan Walker; 148 © Heinrich Volschenk; 150B-R © Thomas Biegalski; 156T-L © Floortje; 160T-R © atılay ünal; 161; 165 © John Cairns; 166B © Paul Tessier; 168L © Franck Boston; 169R © Stephanie Horrocks; 170T-L, 171 © Adam Korzekwa; 170B-L © dra_Schwartz; 185B © muratseyit; 187B © Shaun Lowe; 188C © Pgiam; 188B-R © Duncan Moody; 189T © technotr; 189C © gaspr13; 196B-C, 197T-R © suemack; 196B-R © TebNad; 200BA © Steve Goodwin; 225 © Philip Toy; 235L-C; 237B-L © Vu Banh; 239T-R © pixhook; 245B-R, 246BA, 247BA © Kevin Russ; 246T-L © Simon Askham.

Jane Laurie Illustrations: 114BA; 114B-L; 115BA; 116BA; 117BA; 118BA; 119BA.

Mary Evans Picture Library: 9; 10; 11B © Classic Stock/H. Armstrong Roberts; 14T-R; 17 © AISA Media; 22C-L; 25T; 25B; 27T; 28 © AISA Media; 30T-C; 31T; 31B; 32B; 33; 34; 35L; 36T-C; 36C-L; 37T; 37B; 38; 40T-R; 41T-R; 42T-C; 42C-L; 42B-L; 42L-R; 44C-R; 45L; 45R; 49B; 53C-R; 54T-C; 54B-L; 55; 56T; 57T-L; 57C-R; 57B; 59B © AISA Media; 60T-C; 60B; 61T-R; 61C-R; 61B; 65T-L; 66L; 66R; 68T-C; 68B-L; 69C-R; 69B; 70T, 71C-R © Illustrated London News Ltd; 72B-L; 73T; 73B-R; 74B © Illustrated London News Ltd; 76T-L; 76T-R; 77T; 77B; 78T-L © Illustrated London News Ltd; 78B; 80L, 81, 82T-C © Illustrated London News Ltd; 83B; 84T © Illustrated London News Ltd; 85T © EXPLORER/WOLF; 85B; 88T-L; 88B-L; 89B; 90 © Illustrated London News Ltd; 92 © 94T-C © Blanc Kunstverlag/SZ Photo; 94T-L; 94C-L; 94B-L; 95B-R © Imagno; 96; 97; 103B-R © Classic Stock/H. Armstrong Roberts; 105;T 105B; 108T-C © Rue des Archives/PVDE; 108C-L © Classic Stock/H. Armstrong

Roberts; 118, 119 © Rue des Archives/Tallandie; 120T-C; 121; 122; 124T; 124B; 125T; 128C; 131T-R; 140T-C; 141B; 143C; 150T-L © Classic Stock/H. Armstrong Roberts; 157 © Illustrated London News Ltd; 170T-C; 170T-R; 170C-L; 175B © HENRY GRANT; 216T-R; 224 © Classic Stock/Photo Media; 234T-L; 234B-R; 239B-R; 240T; 240B; 241; 242T-L © Varma/Rue des Archives; 242T-R; 243C-R; 244.
NASA: 158T-R; 232. 233B.
Photo Library: 78T-R; 79B-R.
Science Photo Library: 42T-L © Bo Veisland, MI&I; 58 © Martyn F. Chillmaid; 154B; 158B © Photo Library International; 164T © Michel Viard, Peter Arnold INC.; 176 © A. Barrington Brown; 177T-L; 178 © CNRI; 186B © Dr Jeremy Burgess; 192T-C © Robin Scagell; 199 © Clive Freeman, The Royal Institution; 206T-L © Kenneth Eward; 207B © Joel Arem; 210T-L © P. Dumas/Eurelios; 211 © Geoff Tompkinson; 212BA © Pasieka; 212T-R © David McCarthy; 212B-L © Eye of Science; 213 © Laguna Design; 215T-L © Physics Today Collection/American Institute of Physics; 215C-L; 215B-R © Charles D. Winters; 216T-L © Physics Today Collection/American Institute of Physics; 216C-R; 217L © Takeshi Takahara; 217R © Laguna Design; 218 © Eye of Science; 219, 220C-L © US Department of Energy; 221C-R © Manfred Kage; 227 © CERN; 228T-C © NASA/JPL; 229 © NASA/JPL/Cornell; 230 © NASA; 231 © NASA/JPL-Solar System Visualization Team; 234T-R © Chris Butler; 234B-L © US Geological Survey; 243T; 245T.
3M: 166C-L; 166B-L.
Shutterstock: 12C; 13; 23BA; 24T-L; 25BA; 26T-L; 27BA; 48T-L; 48B-L; 49C-L; 53B; 54T-L; 54B; 56B; 60T-L; 62; 67; 68T-L; 72T-L; 72C-L; 78C-L; 79B; 99; 102T-L; 102B-L; 103T; 104; 105B-R; 106; 107L; 107R; 108T-L; 109B; 110L; 111T-R; 111C-R; 114C-L; 115T; 115B; 116T-C; 117T-L; 117T-C; 117T-R; 120T-L; 120C-L; 120B-R; 121BA; 123BA; 123C; 123B-L; 123B-C; 123B-R; 125B; 126T-L; 126C-L; 126B; 128T; 132R; 134C-L; 134B; 135T; 136; 140BA; 141T; 142; 143T; 146T-L; 146C-L; 146B-L; 147T; 147B; 146T; 150T-R; 154T-L; 154T-R; 154C-L; 155T-R; 155C-L; 156B; 159T-R; 159B; 160C-L; 160B; 166T-L; 167T-L; 168BA; 172; 180T-L; 180C-L; 180B-L; 182T; 182B; 183; 184BA; 184T; 186T-R; 186C-L; 189C-R; 190; 190C; 192T-L; 192C-L; 192B; 193T-L; 193C; 195B; 195BA; 198T-L; 198C-L; 198B-L; 198B-C; 206T-R; 206C-L; 206B-L; 207T-R; 208T-L; 208C; 209B-L; 209B-R; 210B-L; 214T-L; 214T-R; 214TC-L; 214B-R; 214B; 214C-R; 215BA; 217B-R; 220BA; 222T-L; 222C-L; 222B-L; 223B; 226BA; 226B; 228T-L; 228TC-L; 228C-L; 228B-L; 237T-R; 237B-R; 246T-C; 246T-R; 247T-L.
Silvio Tanaka: 222T-C.
Superglue Corporation, California: 160T-L; 162.
Topfoto: 29; 44T-L; 174T; 174B; 175T.

DEDICATION

To Patrick Healey, Adrienne Gaughan, and Rachel Gaughan Blazevic.

ACKNOWLEDGEMENTS

MANY PEOPLE PROVIDED ASSISTANCE.

Thanks to Sheila D'Amico and Adrienne Bischoff, without whose perceptive eyes, insightful comments, and generous spirits the book in your hands would not have been written.

For suggesting potential "accidental geniuses," thanks to Gregory A. Good of the Center for History of Physics at the American Institute of Physics; Barry Shell, founder of The Great Canadian Scientists Research Society; and Patrick Shea, Senior Archivist with the Chemical Heritage Foundation.

Thanks to the Santa Cruz Public Library, the Los Angeles Public Library, and the University of California Libraries at Santa Cruz and Berkeley. The access to information that you provide is appreciated as the invaluable resource it is.

Thanks also to family and friends for their support and inspiration.

INDEX

★